PROBLEMS IN THE PHILOSOPHY OF MATHEMATICS

STUDIES IN LOGIC

AND

THE FOUNDATIONS OF MATHEMATICS

NORTH-HOLLAND PUBLISHING COMPANY
AMSTERDAM · LONDON

PROBLEMS IN THE
PHILOSOPHY OF MATHEMATICS

Proceedings of the International Colloquium in the Philosophy of Science, London, 1965, volume 1

Edited by

IMRE LAKATOS

Reader in Logic, University of London

1972

NORTH-HOLLAND PUBLISHING COMPANY
AMSTERDAM · LONDON

Library of Congress Catalog Card Number 67-20007

ISBN 07204 2240 X

First edition 1967
Second printing 1972

PRINTED IN THE NETHERLANDS

PREFACE

This book constitutes the first volume of the Proceedings of the 1965 International Colloquium in the Philosophy of Science held at Bedford College, Regent's Park, London, from July 11th to 17th 1965. The Colloquium was organised jointly by the British Society for the Philosophy of Science and the London School of Economics and Political Science, under the auspices of the Division of Logic, Methodology and Philosophy of Science of the International Union of History and Philosophy of Science.

The Colloquium and the Proceedings were generously subsidised by the sponsoring institutions, and by the Leverhulme Foundation and the Alfred P. Sloan Foundation.

The members of the Organising Committee were: W. C. Kneale (Chairman), I. Lakatos (Honorary Secretary), J. W. N. Watkins (Honorary Joint Secretary), S. Körner, Sir Karl R. Popper, H. R. Post, and J. O. Wisdom.

The Colloquium was divided into three main sections: *Problems in the Philosophy of Mathematics*, *The Problem of Inductive Logic*, and *Problems in the Philosophy of Science*. The Proceedings are to be published in three volumes, with the same titles.

This first volume, *Problems in the Philosophy of Mathematics*, contains revised, and at times considerably expanded, versions of the nine papers presented in this field at the Colloquium. Some of the participants in the debates were invited to submit discussion notes based on the revised versions of the papers; thus they too differ, sometimes greatly, from the original comments made during the discussions. The authors' replies are in their turn based on these reconstructed discussion notes.

The Editor wishes to thank all the contributors for their kind cooperation. He is also grateful to his collaborators – above all, to Alan Musgrave for his invaluable editorial assistance, and to Miss Phyllis Parker for her conscientious secretarial and organisational help.

THE EDITOR

London, July 1966

PROGRAMME

INTERNATIONAL COLLOQUIUM IN THE PHILOSOPHY OF SCIENCE

London, July 11–17, 1965

July 11, Sunday

p.m.	8.15	W. C. KNEALE: *Presidential welcome*
	8.30	Sir KARL R. POPPER: *Rationality and the search for invariants*

July 12, Monday

PROBLEMS IN THE PHILOSOPHY OF MATHEMATICS I
Chairman: A. TARSKI

a.m.	9.30–10.15	A. ROBINSON: *The metaphysics of the calculus*
	10.15–11.00	*Discussion*
	11.00–12.00	A. MOSTOWSKI: *Recent results in set theory*
	12.00–12.45	*Discussion*

PROBLEMS IN THE PHILOSOPHY OF MATHEMATICS II
Chairman: W. VAN O. QUINE

p.m.	3.00–3.45	P. BERNAYS: *What do some recent results in set theory suggest?*
	3.45–4.30	S. KÖRNER: *On the relevance of post-Gödelian mathematics to philosophy*
	5.00–6.30	*Discussion*

July 13, Tuesday

THE PROBLEM OF INDUCTIVE LOGIC I
Chairman: W. C. KNEALE

a.m.	9.30–10.15	R. JEFFREY: *Probable knowledge*
	10.30–11.15	R. CARNAP: *Inductive logic and inductive intuition*
	11.30–1.00	*Discussion*

PROBLEMS IN THE PHILOSOPHY OF SCIENCE I
Chairman: Sir KARL R. POPPER

p.m.	3.00–4.30	T. S. KUHN and J. W. N. WATKINS: *Criticism and the growth of knowledge*
	5.00–6.30	*Discussion*

PROBLEMS IN THE PHILOSOPHY OF MATHEMATICS III
Chairman: S. C. KLEENE

8.30 G. KREISEL: *Informal rigour and completeness proofs*

July 14, Wednesday (ROOM A)

PROBLEMS IN THE PHILOSOPHY OF SCIENCE II
Chairman: G. J. WHITROW

a.m. 9.30–10.15 L. P. WILLIAMS: *Epistemology and experiment: the case of Michael Faraday*
10.15–10.45 *Discussion*

PROBLEMS IN THE PHILOSOPHY OF SCIENCE III
Chairman: H. BONDI

11.00–11.45 P. G. BERGMANN: *The general theory of relativity – Case study in the unfolding of new physical concepts*
12.00–1.00 *Discussion*

PROBLEMS IN THE PHILOSOPHY OF SCIENCE IV
Chairman: A. J. AYER

p.m. 2.00–2.45 G. MAXWELL: *Scientific methodology and the causal theory of perception*
2.45–3.30 *Discussion*
3.45–4.15 B. JUHOS: *The influence of epistemological analysis on scientific research ('length' and 'time' in the special theory of relativity)*
4.15–5.00 *Discussion*

PROBLEMS IN THE PHILOSOPHY OF MATHEMATICS IV
Chairman: J. S. WILKIE

5.00–5.45 Á. SZABÓ: *Greek dialectic and Euclid's axiomatics*
5.45–6.30 *Discussion*

PROBLEMS IN THE PHILOSOPHY OF MATHEMATICS V
Chairman: G. T. KNEEBONE

8.30 L. KALMÁR: *Foundations of mathematics: Whither now?*

July 14, Wednesday (ROOM B)

THE PROBLEM OF INDUCTIVE LOGIC II
Chairman: S. Toulmin

a.m.	9.30–10.00	H. Freudenthal: *Realistic models of probability*
	10.00–10.45	*Discussion*
		Chairman: J. Hajnal
	11.00–12.00	J. Hintikka: *Induction by enumeration and induction by elimination*
	12.00–1.00	*Discussion*

THE PROBLEM OF INDUCTIVE LOGIC III
Chairman: D. V. Lindley

p.m.	2.00–3.30	H. Kyburg and Y. Bar–Hillel: *The rule of detachment in inductive logic*
	3.45–5.00	*Discussion*

THE PROBLEM OF INDUCTIVE LOGIC IV
Chairman: L. Jonathan Cohen

	5.00–5.45	Mary B. Hesse: *Consilience of inductions*
	5.45–6.30	*Discussion*

July 15, Thursday PROBLEMS IN THE PHILOSOPHY OF SCIENCE V
Chairman: R. Wollheim

a.m.	9.15–10.00	E. Gellner: *Cause and meaning*
	10.00–10.30	*Discussion*
	10.45–11.30	R. A. H. Robson: *The present state of theory in sociology*
	11.30–12.00	*Discussion*
	12.00–12.30	J. O. Wisdom: *Antidualist outlook and social enquiry*
	12.30–1.00	*Discussion*

PROBLEMS IN THE PHILOSOPHY OF SCIENCE VI
Chairman: R. G. D. Allen

p.m.	2.30–3.15	P. Suppes: *Mathematical methods in the social sciences*
	3.15–4.00	*Discussion*
	4.15–5.00	R. D. Luce: *An evaluation of mathematical psychology*
	5.00–5.45	*Discussion*

	5.45–6.15	J. HARSÁNYI: *Individualistic versus functionalistic explanations in the light of game theory: the example of social status*
	6.15–6.45	*Discussion*

PROBLEMS IN THE PHILOSOPHY OF SCIENCE VII

Chairman: T. GORMAN

	8.30	L. HURWICZ: *Mathematical methods in economics*

July 16, Friday (ROOM A)

PROBLEMS IN THE PHILOSOPHY OF SCIENCE VIII

Chairman: D. BOHM

a.m.	9.30–10.00	W. YOURGRAU: *A budget of paradoxes in physics*
	10.00–10.30	*Discussion*
	11.00–11.30	M. BUNGE: *The maturation of science*
	11.30–12.00	*Discussion*

PROBLEMS IN THE PHILOSOPHY OF SCIENCE IX

Chairman: T. J. SMILEY

	12.00–12.30	R. SUSZKO: *Formal logic and the development of knowledge*
	12.30–1.00	*Discussion*

PROBLEMS IN THE PHILOSOPHY OF SCIENCE X

Chairman: B. WILLIAMS

p.m.	3.15–4.00	W. W. BARTLEY: *The rôle of theories of demarcation in the history of the philosophy of science*
	4.15–5.00	*Discussion*
	5.00–5.45	R. POPKIN: *Scepticism, theology and the scientific revolution of the seventeenth century*
	5.45–7.00	*Discussion*

July 16, Friday (ROOM B)

THE PROBLEM OF INDUCTIVE LOGIC V

Chairman: R. B. BRAITHWAITE

a.m.	9.30–11.00	W. C. SALMON and I. HACKING: *The justification of inductive rules of inference*
	11.30–12.30	*Discussion*

		PROBLEMS IN THE PHILOSOPHY OF MATHEMATICS VI
		Chairman: P. SUPPES
p.m.	2.00–2.45	J. EASLEY: *Logic and heuristic in mathematics curriculum reform*
	2.45–3.15	*Discussion*
		Chairman: M. DUMMETT
	3.15–4.00	F. SOMMERS: *On a Fregean dogma*
	4.00–4.45	*Discussion*

July 17, Saturday		THE PROBLEM OF INDUCTIVE LOGIC VI
a.m.	9.30–12.00	*General discussion*

CONTENTS

Preface v

Programme of the colloquium vii

Á. Szabó: GREEK DIALECTIC AND EUCLID'S AXIOMATICS 1

Discussion:

W. C. Kneale: Priority in the use of *reductio ad absurdum* . . 9
L. Kalmár: The Greeks and the excluded third 10
A. Robinson: The Greeks and the excluded third 11
J. R. Lucas: Plato and the axiomatic method 11
P. Bernays: Some doubts about the Eleatic origin of Euclid's axiomatics 14
G. J. Whitrow: The mythical origins of Euclidean geometry . 16
K. R. Popper: The cosmological origins of Euclidean geometry . 18
Á. Szabó: Reply 20

A. Robinson: THE METAPHYSICS OF THE CALCULUS . . 28

Discussion:

P. Geach: Infinity in scholastic philosophy 41
H. Freudenthal: Technique versus metaphysics in the calculus 42
A. Heyting: Technique versus metaphysics in the calculus . . 43
Y. Bar-Hillel: The irrelevance of ontology to mathematics . 44
M. Bunge: Non-standard analysis and the conscience of the physicist 44
A. Robinson: Reply 45

F. Sommers: ON A FREGEAN DOGMA 47

Discussion:

L. Kalmár: Not Fregean and not a dogma 63
M. Dummett: A comment on 'On a Fregean dogma' . . . 63
C. Lejewski: The logical form of singular and general statements 68
W. V. Quine: Three remarks 70
F. Sommers: Reply 71

A. MOSTOWSKI: RECENT RESULTS IN SET THEORY . . . 82

DISCUSSION:

G. KREISEL: Comments 97
A. ROBINSON: Comments 103
L. KALMÁR: On the role of second order theories 104
A. MOSTOWSKI: Reply 105

P. BERNAYS: WHAT DO SOME RECENT RESULTS IN SET THEORY SUGGEST? 109

DISCUSSION:

A. MOSTOWSKI: Cohen's independence proof and second order formalisation 113
Y. BAR-HILLEL: The dangers of Platonistic modes of speech . 114
P. SUPPES: After set theory, what? 115
P. BERNAYS: Reply 116

S. KÖRNER: ON THE RELEVANCE OF POST-GÖDELIAN MATHEMATICS TO PHILOSOPHY 118

DISCUSSION:

G. H. MÜLLER: An old philosophical question – and the recent results in the foundations of mathematics 133
Y. BAR-HILLEL: On a neglected ontology-free philosophy of mathematics 136
S. KÖRNER: Reply 136

G. KREISEL: INFORMAL RIGOUR AND COMPLETENESS PROOFS 138

DISCUSSION:

Y. BAR-HILLEL: Obtaining axioms by reflection 172
A. HEYTING: Informal rigour and intuitionism 172
J. MYHILL: Remarks on continuity and the thinking subject . 173
G. KREISEL: Reply 175

L. KALMÁR: FOUNDATIONS OF MATHEMATICS – WHITHER NOW? 187

DISCUSSION:

A. HEYTING: Weyl on experimental testing of mathematics . . 195
S. C. KLEENE: Empirical mathematics? 195
P. BERNAYS: Mathematics and mental experience 196
Y. BAR-HILLEL: Is mathematical empiricism still alive? . . . 197
I. LAKATOS: A renaissance of empiricism in the recent philosophy of mathematics? 199
L. KALMÁR: Reply 203

J. A. EASLEY, JR: LOGIC AND HEURISTIC IN MATHEMATICS CURRICULUM REFORM 208

DISCUSSION:

P. SUPPES: The central role of empirical knowledge in curriculum reform 231
L. KALMÁR: Mathematics teaching experiments in Hungary . . 233
J. A. EASLEY, JR.: Reply 238

GREEK DIALECTIC AND EUCLID'S AXIOMATICS

ÁRPÁD SZABÓ

Hungarian Academy of Sciences, Budapest

One of the most thrilling, yet hitherto little known, chapters in the history of mathematics is the transformation of early practical and empirical mathematical knowledge into a systematic deductive science based on definitions and axioms [1]. There is no doubt that this highly important transformation took place in ancient Greece. Before the development of Greek culture the concept of *deductive* science was unknown to the Eastern peoples of antiquity. In the mathematical documents which have come down to us from these peoples, there are no *theorems* or *demonstrations*, and the fundamental concepts of *deduction*, *definition*, and *axiom* have not yet been formed. These fundamental concepts made their first appearance only with the Greek mathematicians.

But why did the Greeks not rest satisfied with practical or empirical mathematical knowledge? Why did they replace the existing collection of mathematical prescriptions by a systematic deductive science? What prompted them suddenly to put more trust in what they could prove by theory, or demonstrate, than in what practice showed to be correct? Deductive mathematics is born when knowledge acquired by practice *alone* is no longer accepted as true; when higher-ranking theoretical reasons are needed even for what practice invariably corroborates.

My problem is to explain the change in the criterion of truth in

---

[1] See also my papers: 'Die Grundlagen in der frühgriechischen Mathematik', *Studi Italiani di Filologia Classica*, XXX, 1958, pp. 1–51; 'Der älteste Versuch einer definitorisch-axiomatischen Grundlegung der Mathematik', *Osiris*, XIV, 1962, pp. 308–369; 'The Transformation of Mathematics into Deductive Science and the Beginnings of its Foundation on Definitions and Axioms', *Scripta Mathematica*, XXVII, 1964, pp. 27–49, 113–139; 'Anfänge des Euklidischen Axiomensystems', *Archives for History of Exact Sciences*, I, 1960, pp. 37–106.

mathematics from justification by practice or experience to justification by theoretical reasons. My solution is that this change was due to the impact of philosophy, and more precisely of Eleatic dialectic, upon mathematics. In my short lecture I shall try to explain how Greek deductive mathematics was founded by adapting Eleatic dialectic to mathematical knowledge.

My lecture has three parts. In *part 1* I shall explain the connections between dialectic and the organization of Euclid's *Elements*. In *part 2* I shall discuss the so-called indirect proof. Finally, in *part 3*, I shall explain how one of Euclid's axioms was formulated in order to circumvent one of Zeno's paradoxes.

1. Euclid begins the *Elements* by listing three kinds of mathematical principles, the *definitions*, the *postulates*, and the *axioms*. In the fifth century A.D., Proclus, the neo-Platonic scholiast of the *Elements*, dwells in detail on the meaning and significance of these unproved principles for the whole of mathematics. He says[1]:

> Since we assert that this science, geometry [i.e. mathematics] starts from suppositions and proves its statements by definite principles . . . whoever compiles a manual of geometry has to treat separately the bases of this science, and separately the conclusions that he deduces from the principles. He has not to give account of the principles, he is not bound to prove them, but he has to prove everything deduced from the principles.

Proclus makes it clear elsewhere that the unproved suppositions of mathematics (in Greek, the ὑποθέσεις) are the *definitions*, *postulates* and *axioms*. In the quoted passage Proclus asserts that mathematics is a *hypothetical science*, i.e. that in mathematics we start from suppositions or principles which we do not prove, but accept as true without proof, and prove only the theorems deduced from the principles.

Two questions now arise. First, how long had Greek mathematicians known that their science was a hypothetical one, that they

---

[1] Proclus, *In Euclidem Comm.*, G. Friedlein ed., 1873, pp. 75–76 *et seq.*

had to start from statements that did not have to be proved? Second, how did mathematicians discover that their first principles were not in fact to be proved?

An answer to the first question is easily found if we remember what Socrates says about the mathematicians in Plato's *Republic* [1]:

> You know, to be sure, that those who deal with geometry, arithmetic and similar sciences, take the *odd*, the *even*, the *figures*, the three sorts of *angles* and other such things as bases. These are the starting points for their investigations, and they do not believe that they are compelled to give any account of them. That is where they start from as principles.

This passage shows that in the age of Plato mathematicians were already clear that they do not demonstrate their principles, but accept them as true without proof. The passage also shows that the listed concepts – the *even*, the *odd*, the geometrical *figures* and the three sorts of *angles* – are the so-called *hypotheses* of mathematics, so that mathematics, in containing such hypotheses, is a *hypothetical* science.

The Greek word ὑπόθεσις derives from the preposition ὑπό and the verb τίθεσθαι, and signifies, in fact, that which two conversationalists, the partners in a debate, mutually agree to accept as the basis and starting point of their debate. Thus the Greek ὑπόθεσις is not identical with our 'hypothesis'. What we call hypothesis is a theory which we have not yet proved, but which must be proved later. On the other hand, the Greek ὑπόθεσις in dialectic and mathematics is a starting point which it is impossible to prove; and it does not *need* to be proved, because the partners in the debate agree upon it. Since the debating partners both agreed about what *odd* and *even* mean, and what the geometrical *figures* are, and what are the three sorts of *angles*, these are the unproved *hypotheses* of mathematics.

These considerations about the term ὑπόθεσις provide the answer to our second question: how had the Greek mathematicians discovered that the starting principles of mathematics could not be,

---

[1] *Republic*, VI, 510 C–D.

and did not need to be, proved? They seem to have come to this from the practice of dialectic. They were accustomed to the fact that, when one of the partners in a debate wanted to prove something to the other, he was bound to start from an assertion accepted as true by both of them. Such an assertion, accepted by both partners, was called *ὑπόθεσις* – the ground of the debate. This method was retained in systematic mathematics also, which was based upon statements believed to be accepted by everyone without proof, and called the hypotheses of mathematics. The first kind of hypotheses are the *definitions* which for the Greeks were circumscriptions, given without proof, of the concepts (notions) used in mathematics.

2. In this second part of my lecture I shall argue that the most interesting method of proof in Greek mathematics, the *indirect proof*, derives from the dialectic of the Eleatics.

In order to understand the idea of indirect proof we must first see how the 'hypotheses' were used in dialectic and mathematics. Socrates, in the Platonic dialogue *Phaedo*, characterizes his method as follows [1]:

> I always start in my thinking from an assertion judged by me to be strongest. This assertion is my *hypothesis*; and what seems to harmonize with this assertion, I consider to be true. If I see, however, that something is not in harmony with the previous strong assertion, I look upon it as untrue.

I would like to draw attention to two points in this passage:

(1) As we see, Socrates uses his hypotheses as starting points, as is done in dialectic and in mathematics.

(2) Socrates wants to find what is in harmony with his starting point, and what is not. But how did Socrates determine that an assertion was in harmony with another assertion chosen as starting point? The answer to this question is that this was just the purpose of demonstration or proof.

To explain this I have to modify, or rather expand, what I have

---

[1] *Phaedo*, 100 A.

said so far about the meaning of ὑπόθεσις or 'hypothesis' in dialectic [1]. I said above that the Greeks meant by *hypothesis* a strong initial assertion, which was never proved, but accepted as true without proof – and this is indeed its most frequent meaning. But sometimes the same term means a tentative assertion, made so that we can investigate its truth. This second meaning is almost the same as our meaning of 'hypothesis'. And with this in mind, we can understand the idea of indirect demonstration, an example of which I shall now present.

In Plato's *Theaetetus* [2] the participants want to settle the problem of whether or not knowledge and sensory perception are identical. To do this they first suppose that '*knowledge and sensory perception are identical*'. This statement is the *hypothesis* of their investigation, and they first investigate what results from this hypothesis.

If knowledge and sensory perception are identical, then seeing is identical with knowing, because seeing is one sort of sensory perception. This means that a person who *sees* something also *knows* what he sees, and a person who *does not see* something *does not know* what he does not see. Now suppose a person sees something, and thus, according to the hypothesis, knows it also, and then shuts his eyes. When he has closed his eyes he does not see the thing, and so, according to the hypothesis, cannot know the thing either. But clearly the person will know, when his eyes are shut, what he had seen when his eyes were open. Thus our hypothesis leads us to say that the person does not know the thing, and that he does know it. The assumption that knowledge and sensory perception are identical leads to a *contradiction*.

As Socrates puts it: something impossible results from the assumption that knowledge and sensory perception are identical; this hypothesis implies a *contradiction*, and therefore it cannot be true. Therefore the opposite of our hypothesis must be true: '*knowledge is not identical with sensory perception*'.

This example shows the two peculiarities of indirect demonstration:

---

[1] See also my paper 'Anfänge, etc.' ch. 8.

[2] *Theaetetus*, 163 A–B.

(1) In an indirect demonstration we do not try to demonstrate that assertion which we guess to be true, but we try rather to *refute* the opposite assertion. But to refute the opposite assertion is in effect to demonstrate the assertion we are interested in, because either A is true or its opposite is true, and there is no third possibility. If we refute one of two contrary assertions, the other must be true.

(2) In an indirect demonstration we refute an assertion by showing that it leads to a contradiction. The refutation consists in making evident the contradiction.

We have seen an example of the use of indirect proof in dialectic. But Plato many times calls indirect proof a special form of *mathematical* proof [1], and you will remember that it is often to be found in Euclid. Moreover, as is widely known, according to Aristotle the Pythagoreans, as early as the fifth century B.C., had demonstrated the incommensurability of the diagonal and the side of the square by pointing out that the opposite assertion – 'The diagonal and the side of the square *are* commensurable quantities' – cannot be true since it leads to the contradiction, 'The same number is odd and even at the same time' [2], which is absurd (*ἀδύνατον* [3]).

It seems clear that this special form of mathematical demonstration was originally used in philosophy. For Eleatic philosophy, the doctrines of Parmenides and of Zeno, rested upon indirect demonstrations which revealed the contradictions in propositions opposite to their own. Thus I have argued [4] that Greek mathematicians took the idea of indirect demonstration from Eleatic philosophy.

3. Having argued in the first part of my lecture that Euclid's mathematical 'principles' were an adaptation of the dialectician's 'hypotheses', and in the second part the origin in dialectic of the indirect demonstration, I shall now show how one of the axioms

---

[1] See my paper 'Anfänge etc.', part II, chs. 5 and 6.
[2] Cf. T. H. Heath, *Mathematics in Aristotle*, Oxford, 1949, pp. 2, 22, etc.
[3] See my paper 'Anfänge etc.', ch. 9.
[4] In the papers cited in the first footnote.

in Euclid's *Elements* was included simply as an answer to one of Zeno's paradoxes.

The axiom in question is: 'The whole is greater than the part' [1].

This axiom is, at first sight, obvious and indubitable. So why did it occur to anyone to begin by stating it, as a strong assertion, without proof? Could the reason why this seemingly obvious proposition was stated as an axiom be that it had actually been called into question by somebody?

A text of Aristotle makes it clear that one of Zeno's paradoxes was: '*Half the time is equal to its double*' [2]. It seems to me that this paradox asserts the very opposite of Euclid's axiom stated above.

What can we make of Zeno's strange paradox? Aristotle and his disciples thought it a mere sophism. And, it must be remembered, we know of Zeno's argument only from Aristotle's account. I feel that Aristotle's judgement of the paradox is unjust to Zeno. It is likely that Zeno could have argued for his paradox, because for him the 'whole time' as well as 'half the time' were both infinite sets. And these two infinite sets, if not entirely 'equal', are equivalent, in the modern set-theoretical sense.

In a sense, therefore, Zeno was right: his assertion that '*Half the time is equal to its double*' could not have been refuted. This explains why Euclid was compelled to assert, as an axiom, that 'The whole is greater than the part'. For he wanted to confine himself to finite sets, while Zeno considered infinite ones.

This provides us with an answer to the question, 'What is the role of the axioms in Greek mathematics?'

In the first part of my lecture I explained how in dialectic the two partners in a debate tried to proceed from an assertion accepted and believed to be true by both of them. This mutually accepted assertion was called the 'hypothesis' of the debate.

But what happens if the disputants can find no mutually acceptable assertion from which to begin? What happens if, for example, one of them asserts that '*Half the time is equal to its*

---

[1] Euclid, *Elements*, κοιναὶ 'ἔννοιαι VIII. In our text of Euclid ἀξιώματα are called κοιναὶ 'ἔννοιαι: but see Part I (Die Terminologie des Proklus und unser Euklid-Text) of my paper 'Anfänge etc.'.

[2] H. Diels – W. Kranz, *Fragmente der Vorsokratiker*, I[8] 29 Zenon A 28.

double' while the other asserts that '*The whole is greater than the part*'? In this case there is no mutually acceptable basis for the subsequent discussion; and one of the partners cannot begin from a 'hypothesis' but only from a strong assertion taken as starting point *by him* – from an '*axiom*'. The Greek word 'axioma' originally meant only '*request*'; one partner *requested* the other to accept his assertion as the starting point of the debate. Euclid's much-discussed axiom, '*The whole is greater than the part*', was also such a request.

I hope that my short lecture has been able to point out, at least in general outline, the close historical connections between Euclidean Axiomatics and Eleatic dialectic.

DISCUSSION

W. C. KNEALE: *Priority in the use of reductio ad absurdum.*

I should like to say first how much I have enjoyed Professor Szabó's paper. There is, however, one question I wish to raise. and that is about the order of priority between the use of *reductio ad absurdum* in metaphysics and its use in mathematics. I do not think we should assume that mathematicians cannot use a logically valid pattern of reasoning in their work until some philosopher has written about it and told them that it is valid. In fact we know that this is not the way in which the two studies, logic and mathematics, are related. Obviously men can recognize some arguments as valid and others as invalid before they have any general theory of validity.

Now it seems probable to me that the procedure of *reductio ad absurdum* was used in mathematics before it was used by Zeno the Eleatic in development of the peculiar metaphysical theory he shared with Parmenides. When Aristotle writes about *reductio ad absurdum*, he quotes as his standard example the proof of the incommensurability of the diagonal with the side of a square, and I think it likely that an early version of that proof (perhaps not as tidy as the one now printed in the Appendix to Heiberg's *Euclid*) was known among the Pythagoreans for some time before Zeno produced the paradoxes which made him famous. It is true that our history of the Pythagoreans is guess-work. We cannot be certain just exactly what they discovered and when, nor yet indeed who they were. But we have a legend (and legends sometimes contain scraps of truth) that a certain Hippasus of Metapontum was expelled from the order, and according to one version drowned at sea, for revealing to the uninitiated. in defiance of a rule of secrecy. that there are incommensurable magnitudes. This suggests that the absence of a ratio in the strict sense between the diagonal and the side of a square had already been discovered by some Pythagorean but was considered so shocking by members of the order, because of its incompatibility with their view about the importance of ratio (logos) in the structure of the world, that they decided to keep it secret from the general public until they had found some way of coping with the difficulty it presented. This interpretation gives some sense at least to the sad story of Hippasus; and although we cannot be certain of his date, I think we may safely assume that he was earlier than Zeno. If so, when Zeno used *reductio ad absurdum* in his queer arguments against the possibility of motion, his reason may have been that this pattern of argument had already had a great success of a negative kind. If attention had been drawn to it because of a really impressive achievement of disproof in which it had played a part, philosophers might well have begun to think of it as a method to be developed for metaphysical purposes.

It is true that Aristotle speaks of Zeno as the inventor of dialectic. But I think that in this passage he must mean by 'dialectic' the metaphysical use of the method of *reductio ad absurdum*, rather than the method itself. For it does not seem plausible to suppose that the famous result about the incommensurability of the diagonal with the side of a square was not achieved until after Zeno had produced his paradoxes.

L. KALMÁR: *The Greeks and the excluded third.*

Professor Szabó claims that the basis of indirect proof in ancient Greek mathematics was the principle of the excluded third. Now mathematicians usually call different kinds of reasonings 'indirect proof'. There are two main kinds, one of which, viz. proof of a theorem p by assuming its negation $\neg p$ and deriving from it a proposition q whose negation has already been proved, requires the principle of the excluded third, even according to the standards of modern mathematical logic. (Indeed, the false consequence q of $\neg p$ shows nothing but that $\neg p$ cannot hold, which implies p only if we assume that either p or $\neg p$ must hold.) On the other hand, the second kind of 'indirect proof', viz. disproof of an assertion p by assuming it and deriving from p a proposition q whose negation has already been proved, does not require, according to modern mathematical logic, the principle of the excluded third. (Indeed, the false consequence q of p shows that p cannot hold and $\neg p$ expresses just this fact; hence, $\neg p$ is proved, i.e. p is disproved without using the principle of the excluded third.) All the examples given by Szabó are of the second kind of 'indirect proof', i.e. of the kind not requiring the principle of the excluded third. Of course, it would be quite unhistorical to require of the ancient Greeks the fine distinctions which are made today in mathematical logic, but the question arises whether all 'indirect proofs' of the ancient Greeks were of the second kind or whether they used also indirect reasonings which require the principle of the excluded third, hence, indirect reasonings not accepted by contemporary intuitionists.

A. ROBINSON: *The Greeks and the excluded third.*

Referring to a matter which has been raised by Professor Kalmár, there is no doubt that the Greeks were familiar with the method of proof by *reductio ad absurdum.* In particular – as Professor Kneale already mentioned – the familiar argument for showing that the square root of 2 is irrational goes back to them. But there is also a second kind of negative proof in Euclid. Thus, when Euclid shows that every composite number has a prime divisor he uses the method of infinite descent, concluding 'which is impossible for numbers', or words to that effect. A contemporary intuitionist would accept the negative proof in the first example since it is a proof of a negative fact (that there are no positive integers m and n such that $m^2=2n^2$), while rejecting the second proof, as it stands, since it tries to establish the proof of a positive, existential, statement. It would be interesting to know whether the Greeks were aware of this distinction.

J. R. LUCAS: *Plato and the axiomatic method.*

I also would very much like to agree with all that Professor Szabó has said and I am only going to ask whether he would accept a slight change of emphasis; Szabó gives most of the credit to the

Eleatic school, but I think that even on the evidence he cited, one can enter more of a plea for Plato. The Eleatics certainly produced the argument by *reductio ad absurdum*: but it seems to me that it is Plato who put forward the ideal of axiomatization as something to be pursued fairly consciously. The seventh book of the *Republic* reads to me rather like the programme of studies for the Institute of Advanced Studies at Athens; in fact one of his pupils was Eudoxus, who did carry through his programme very far. The other reason why I think that one should give Plato the credit is that it does become fairly intelligible why Plato, at the time when he was writing the sixth and seventh books of the *Republic*, should be anxious to move from the rather simple, informal dialectic – the form of argument between two people with some sand where you draw some diagrams and you see that some three lines must be concurrent – to the idea of having definite *ὑποθέσεις* (hypotheseis) from which certain things follow deductively by *διάνοια* (dianoia). I think there are two reasons for this. First of all, Plato was at this time being very self-conscious about his method and was beginning to discover what it was to be a deductive argument, and in the *Phaedo* passages Szabó quotes[1] you can see the first attempt, and in the *Republic* you can see a second attempt to produce a methodology of argument which is becoming more formal. The second reason why Plato would make this move is that whereas as a young man he was a very enthusiastic *arguer*, as an old man, partly as a result of arguing with unpleasant characters like Thrasymachus, he becomes very much a *lecturer*. The early dialogues *are* dialogues: the middle dialogues are glorified seminars – there is the king-pin and there are some yes-ful youths who say, Of course *Πῶς γαϱ οὔ*; or *Τί μήν*;–. By the time you get to the *Laws* it is simply straight lecturing, yard after yard of it. This change was certainly going on in Plato's general philosophy and in his style, both of which change from being a genuine dialogue to being a monologue. And so similarly, he is trying to make his theory of mathematics (not only mathematics, but mathematics is clearly the example he has in mind in the *Republic*) apply to

---

1 100A, 101D–E.

more than merely *one* particular context of discussion. Professor Szabó rightly said a hypothesis is an assumption which we make for the present purpose; but Plato in the *Phaedo* and *Republic* VI and VII is trying to make his assumptions ones which do not have to be taken for granted for the present, particular case; he is trying to make them ones which must be conceded by everyone. This is the search for the *ἀνυπόθετον ἀρχήν* (anupotheton archen), the fundamental axiom which you do not have to ask someone to grant; it is something which must be conceded by anyone, even by a sophist. It is for this reason that Plato is putting forward the axiomatic ideal, that we should try and develop the whole of our mathematics by deductive reasoning, *διάνοια* (dianoia), from some principles which he (wrongly) thought could be established beyond all possible question. Plato put forward this programme. His pupils largely carried it out. We have the final result, as codified by Euclid. This is my chief argument which is not really a disagreement with Professor Szabó, but a slight change of emphasis.

Let me too finally say a few words about the problem of the status of indirect proofs. I think it is much easier to see how they were readily accepted in Greek, in the context of conversation, dialogues, than they are by modern mathematicians who, thanks to Plato, feel that mathematical arguments ought to be in the form of proof sequences from a set of axioms according to a finite set of rules of inference. With proof sequences, all you require of the word *not* is that it shall not give you inconsistency; the way into the concept of negation in the axiomatic approach to mathematics is by the concept of inconsistency. Whereas if the word *not* obtains its meaning from the context of a dialogue, an argument between two people who are actually there, then we start not with the word *not* so much as the word *no*, and the fundamental idea is that of contradiction. 'Do you agree, Thrasymachus', said Socrates, or 'Do you agree', said Zeno to his victim, 'that this line has so many points?'. And if he says no, then the first arguer will show that this is wrong, and if the answer *No* has been shown to be wrong then it is natural to say that he has established his original case. That is, leaving aside the intuitionist's query that *not not* equals *yes*, it is certainly so that No, No – if one person says 'No' and then, retracting, adds 'That's wrong', then it does seem, in the context of an ordinary dialogue, to have the idea of 'Yes'. To put it another way: it is perfectly possible for an intuitionist dealing with an axiomatic system to take objection to the law of the excluded middle. But if I am arguing in Greece, I can very reasonably ask a person which horn of the dilemma he will take, and if he won't say either p or not p, it's equivalent to having resigned from the game. Now I think this is only a part answer to Professor Kalmár's case, but I think it's one of the reasons why the Greeks were much happier about the law of double negation and the argument of *reductio ad absurdum* than Professor Szabó feels they ought to have been. The answer is that they got this idea in the days when dialectic really was a dialogue between two persons, and it is only thanks to Plato and the ideal of axiomatization that we have come much more to the position of a monologue where these truths seem no longer to be necessary ones.

P. BERNAYS: *Some doubts about the Eleatic origin of Euclid's axiomatics.*

I should like to begin by raising some doubts about Szabó's thesis that the main ideas of the deductive, axiomatic method of Greek geometry were borrowed from Eleatic dialectic. First, it is possible that the method of deductive proof was known *before* the Eleatics, for example by Thales. We are told that Thales knew that the angle subtended at the circumference by the diameter of a circle is a right-angle. It is difficult to see how he could know this, unless he had given some sort of deductive proof of it.

Szabó's main argument for his thesis is that the terminology of the geometers is the same as that used in dialectic. But although the *terminology* of dialectic may have been adopted by the geometers, this does not show that the *methods* were the same, for as we all know, terminology changes. Thus the geometers may have taken their *terminology* from dialectic, and their *methods* from earlier mathematicians.

According to Szabó, the idea of beginning with unproved initial assumptions was borrowed by the geometers from dialectic. But even admitting this, it does not explain the puzzling fact that Euclid has *three kinds* of initial assumptions – the *ὅροι* (definitions), the *αἰτήματα* (axioms), and the *κοιναὶ ἔννοιαι* (common notions). Certainly there is an epistemological distinction between them. Now was this epistemological distinction borrowed from dialectic, as Szabó would have to maintain?

In this connection, it is very queer that some of the 'definitions' are not used in the proofs at all, so that they are certainly not *starting-points* of the proofs, as were the 'initial assumptions' of dialectic. Some researchers explain this by saying that these definitions did not originate with Euclid, but rather were later additions. As to the problem of the difference between the 'axioms' and the 'common notions', perhaps 'axioms' are problematic starting-points, and the 'common notions' unproblematic starting-points, concerning the general laws of *quantities*. It may well be that the programme of Euclidean geometry was to understand the geometrical relations upon the basis of the laws of quantities. The attitude about this was that the axioms on quantities were explicitly stated, whereas it was *silently* assumed that lengths, angles and areas are quantities.

By the way, it is often stated that we find in Euclid an underlying idea of a primacy of geometry over arithmetic, and a tendency to restrict the application of the number concept. But this is hardly true: indeed the theory of proportions in Euclid makes essential use of the idea of number. We know from Hilbert that a purely geometrical theory of proportions can be developed; but the Greeks did not do this.

Let me ask another question. We now know that there is no essential distinction between the things we prove and the things we assume without proof; for we can choose different sets of axioms. Did the Greeks know this? We should not simply assume that they did not: they were probably cleverer than we might think. But would this then not be an argument for a major difference between Eleatic dialectic and Euclidean axiomatics?

Having questioned and criticized some of Dr. Szabó's views, I would like to point out that he has elsewhere solved, in an interesting way, some problems about Euclidean geometry with the help of reference to the influence of Eleatic philosophy. For example, it is very puzzling that in some places we have in Euclid great care and subtlety in the proofs, and in others we find that axioms essential to the proofs are not even stated. For example, in the theory of congruence, the fundamental theorem on congruent triangles is proved by *moving* one triangle so that it rests on top of the other. But no axiom about moving triangles is laid down at the start. Szabó explains this as follows: the Eleatics, with their paradoxes of motion, had shocked mathematicians, and the mathematicians wanted to eliminate motion from mathematics (just as modern mathematicians are shocked by set-theoretical antinomies, and want to eliminate them). Euclid found that he had to introduce motion in this one place, but he did it surreptitiously and was afraid to state explicitly an *axiom* on motion.

G. J. WHITROW: *The mythical origins of Euclidean geometry.*

I should like to congratulate Professor Szabó very much on this extremely interesting and stimulating paper. I have read some of his papers that I have seen in print in journals in the last two years or so, and it was with great eagerness that I looked forward to his talk this afternoon and I am certainly not disappointed.

I agree with pretty well everything he has said but I would like to add something because there is an important aspect of this problem that has puzzled most of us ever since we came across Euclidean geometry at school. How was it that anybody ever came to develop this subject? It is obvious in a way why people should invent arithmetic and mensuration – but why should anybody ever have developed Euclidean geometry?

I would just like to mention something that I think complements what has been said concerning the role of dialectic and the influence of people like Parmenides and so on, leading down to Euclid. I refer to the work of Seidenberg [1], who has written on the ritual

[1] A. Seidenberg, *The Diffusion of Courting Practices*, Univ. of California Publ. Meth. 3 1960; also his paper in the *Archives for History of Exact Sciences*, I, No. 5, 1962, pp. 488–527.

origin of arithmetic and geometry and has been influenced very much by a man who I think has not always received the credit that is his due, the late Lord Raglan. He was, as I expect most of you know, the author of a very interesting theory on the ritual origin of many aspects of civilization. And Seidenberg has applied this general idea of Lord Raglan's to the origins of mathematics, in particular the origin of geometry and even of arithmetic, but I won't say anything about that.

Now, in Greek mathematics, when we are thinking of Euclid, we usually think of theorems, but there are also problems. Many of the problems in Greek geometry I think are of very great interest. I will just mention one in particular that everybody has heard about, the problem of the duplication of the cube. Now you may remember this problem had a mythical origin in connection with a cubical altar to a Greek god. The god was supposed to have said that this altar was to be doubled in size and first one was made of twice the length, and twice the breadth and twice the depth, so of course it was eight times the size of the original, and then the plague returned and so they had to think again and work out some construction whereby they could get a cube twice the volume of the original one. Obviously the origin of this problem, according to the legend, is not to do with dialectic but is to do with a problem of ritual and arises in connection with religion. I think that there are other instances too of course, where there are ritual origins to problems, for example, the problem of making a square equal in area to a circle, the problem of squaring the circle. Now why should anyone want to do that? For all practical purposes this could be done without any exact knowledge of π – of course in antiquity nobody got to an exact numerical value of π, although Archimedes made a fairly good shot at it. But the problem was not just what square is more or less equal to a circle, but what square is exactly equal to a circle – and why should any normal person worry about this if it is not for some sort of ritual reason?

Just as we have this very interesting development from dialectic leading up to the work that has been spoken about by Szabó, so also there is an equally important but I think comparatively neglected line of approach from ritual that influenced the Greeks which should be taken into account when considering the question of the origin of exact mathematics, as distinct from approximate mathematics.

K. R. POPPER: *The cosmological origins of Euclidean geometry.*

(1) I should like first to say how much I too enjoyed Professor Szabó's wonderful paper. His thesis, that the axiomatic method of Euclidean geometry was borrowed from the methods of argument employed by the Eleatic philosophers, is an extremely interesting and original one. Of course, his thesis is highly conjectural, as must be any such thesis, in view of the scanty information that has come down to us about the origins of Greek science.

(2) Szabó, it seems to me, has only explained one facet of Euclidean geometry, how the *method* employed in it was invented. The question to which he has offered a tentative answer is: 'How did Euclid come to adopt the *axiomatic method* in his geometry?' However, I wish to suggest that there is a second, perhaps more fundamental, question. It is this: 'What was the *problem* of Euclidean geometry?' Or to put it another way: why was it *geometry* that was developed so systematically by Euclid?

(3) These two questions are different, but, I believe, closely connected. I should like just to mention a historical conjecture of my own about this second problem [1]. It is this: Euclidean geometry is not a treatise on abstract, axiomatic mathematics, but rather a treatise on cosmology; that it was proposed to solve a problem which had arisen in cosmology, the problem posed by the discovery of irrationals. That geometry was the theory dealing with the irrationals (as opposed to arithmetic, which deals with 'the odd and the even') is repeatedly stated by Aristotle [2].

The discovery of irrational numbers destroyed the Pythagorean programme of deriving cosmology (and geometry) from the arithme-

[1] The conjecture is stated more fully in 'The Nature of Philosophical Problems and their Roots in Science', *The British Journal for the Philosophy of Science* 3 1952; reprinted with corrections as ch. 2 of *Conjectures and Refutations*, 1963, 1965.

[2] For references see *Conjectures and Refutations*, p. 87, note 42.

tic of natural numbers. Plato realised this fact, and sought to replace the arithmetical theory of the world by a geometrical theory of the world. The famous inscription over the gates of his Academy meant exactly what it said: that arithmetic was not enough, and that geometry was the fundamental science. His *Timaeus* contains, as opposed to the previous arithmetical atomism, a *geometrical* atomic theory in which the fundamental particles were all constructed out of two triangles which had as sides the (irrational) square roots of two and three. Plato bequeathed his problem to his successors, and they solved it. Euclid's *Elements* fulfilled Plato's programme, since in it geometry is developed autonomously, that is, without the 'arithmetical' assumption of commensurability or rationality. Plato's largely cosmological problems were solved so successfully by Euclid that they were forgotten: the *Elements* is regarded as the first textbook of pure deductive mathematics, instead of the cosmological treatise which I believe it to have been.

As to Professor Szabó's problem, why the *axiomatic method* was first employed by Euclid, I think that an analysis of the cosmological prehistory of Euclidean geometry may also help to solve this problem. For the methods of solving problems are frequently inherited with those problems. The pre-Socratics were trying to solve cosmological problems, and in so doing, they invented the critical method, and applied it to their speculations. Parmenides, who was one of the greatest cosmologists, used this method in developing what may perhaps have been the first deductive system. One may even call it the first deductive physical theory (or the last pre-physical theory before that of the atomists whose theory originated with a refutation of Parmenides' theory; with the refutation, more especially, of Parmenides' conclusion that motion is impossible, since the world is full [1]).

None of this is inconsistent with the views of Szabó who finds the origins of the deductive method in the Eleatic method of dialectic, or of critical debate. But the link with cosmology adds an extra dimension, in my view a necessary one, to his discussion.

---

1 *Op. cit.*, pp. 79–83.

For it seems to me that the sharp distinction, on the basis of their different *methods*, between mathematics and the natural sciences would have been foreign to the Greeks. Indeed, it was the remarkable success of Euclid which brought about this distinction in the first place. For up to (and, in my opinion, including) Euclid, Greek mathematics and Greek cosmology were one or very nearly so. To understand fully the discovery of '*mathematical methods*', we have to remember the *cosmological problems* which they were trying to solve using these methods. Parmenides was a cosmologist; and it was in support of Parmenides' cosmology that Zeno developed his arguments which, as Professor Szabó stresses, inaugurated the specific Greek way of mathematical thought.

A. SZABÓ: *Reply.*

First of all I should like to express my thanks to all those who contributed, by their criticisms, comments and questions, to the further clarification of the problems I discussed in my paper. I should like to divide my reply into two parts. First, I should like to discuss those problems and questions to which I hope to be able to give a clear-cut answer. Second, I should like to discuss those where I feel unable to do this, and expect a reply only from future research.

(1) During the discussion the problem of the mythical origin of Euclidean geometry was mentioned (by Dr. Whitrow). I can very well imagine that such a viewpoint may be quite fertile for future research into the origins of Greek mathematics. Moreover, we can point out immediately that in the beginning Greek mathematics – not only among the Pythagoreans but even for Plato – belonged *to the sphere of religion*. Let us not forget that according to Plato mathematics – whether arithmetic or geometry – is not concerned with the changing world but with something that is unchanging and eternal. The objects of mathematics belong, like unchanging and eternally true objects, to the sphere of the divine. In this respect the idea that deductive mathematics should have *mythical origins* seems an *ab ovo* plausible idea. However I should

like to stress that in my paper I wanted to restrict my discussion to the *immediate origin of the axiomatic method* of Euclidean geometry. This is why I did not discuss its more distant connections with religion and myth.

(2) I also agree with the other important comment which mentioned the connection of deductive mathematics with *cosmology* (Sir Karl Popper). I do not think that Euclidean geometry can be just a 'cosmological dissertation' which aimed at solving the problem of irrationality as it appeared in cosmology ('Euclidean geometry is not a treatise on abstract, axiomatic mathematics, but rather *a treatise on cosmology* . . . it was proposed to solve a problem which had arisen in cosmology, the problem posed by the discovery of irrationals'). But I myself have emphasized in another connection [1] that the philosophy of Parmenides – in fact the whole Eleatic dialectic and so the whole problem of indirect proof – can be most easily explained historically as being a criticism of Anaximenes' cosmogony. Since deductive mathematics would, I think, never have been developed without Eleatic dialectic, I also think that the development of the Euclidean method must have in fact been closely connected with the much more ancient Greek cosmology.

(3) Professor Bernays raised some doubts about my thesis that the main ideas of the deductive, axiomatic method of Greek geometry were borrowed from Eleatic dialectic. He said: 'it is possible that the method of deductive proof was known *before* the Eleatics, for example by Thales. We are told that Thales knew that the angle subtended at the circumference by the diameter of a circle is a right-angle. It is difficult to see how he could know this, unless he had given some sort of deductive proof of it.' In connection with this argument I should like to draw attention to the following. Unfortunately in our sources there are no data available at all concerning the mathematics of Thales' mathematical proof. I certainly assume that Greek mathematicians tried to prove their assertions already before the Eleatic period. But we can only

[1] 'Zum Verständnis der Eleaten', *Acta Antiqua Academiae Scientiarum Hungaricae* II, 1954, Budapest, pp. 247–254.

guess what these most ancient proofs could possibly be like. In an earlier paper of mine [1] I tried, starting from the Greek technical term for mathematical proof, the verb *δείκνυμι* (*deiknymi*), to reconstruct a more ancient intuitive mathematical proof form which still was close to the empirical. The original etymological meaning of *deiknymi* certainly indicates that old Greek mathematical proofs were not so much deductive but rather empirico-intuitive. Most probably Thales' proofs had this character. I should like to mention here that Imre Lakatos in his 'Proofs and Refutations' discusses very convincingly the problem of informal thought experiment-proof from which only later criticism crystallizes out a deductive structure [2].

(4) Another interesting argument in Professor Bernays' comment to which I should like to refer is this: 'although the *terminology* of dialectic may have been adopted by the geometers, this does not show that the *methods* were the same, for as we all know, terminology changes. Thus the geometers may have taken their *terminology* from dialectic, and their *methods* from earlier mathematicians'. I think that in this particular case we need not separate terminology and method. Not only the terminology but also the method of Greek mathematical proof is identical with the terminology and method of dialectic. Since we cannot explain the emergence of this terminology and method within mathematics but can explain very easily and in fact from step to step the emergence of the same method and terminology within dialectic, it is reasonable to assume that this mathematical method and terminology came from dialectic.

(5) Professor Bernays also stressed: 'Even admitting that the idea of beginning with unproved initial assumptions was borrowed by the geometers from dialectic, this does not explain the puzzling fact that Euclid has *three kinds* of initial assumptions – the *ὅροι* (definitions), the *αἰτήματα* (postulates), and the *κοιναὶ ἔννοιαι*

---

[1] '*Deiknymi*, als mathematischer Terminus für *beweisen*', *Maia* N.S. x, 1958, pp. 106–131.

[2] I. Lakatos, 'Proofs and Refutations', *The British Journal for the Philosophy of Science* **14,** Nos. 53, 54, 55, 56, 1963–64.

(common notions). Certainly there is an epistemological distinction between them. Now was this epistemological distinction borrowed from dialectic?' I have to admit that the brief sketch of this problem which I gave in my paper does not shed sufficient light on the question why Euclid starts the discussion of geometry with the *three* groups of unproved assumptions. It would take us very far if I tried now to discuss this problem in my reply. However since it has been mentioned, let me perhaps refer to a longish paper of mine [1], a few ideas of which I summarised; one chapter of this paper discusses exactly this problem: *Euklids dreifache Unterscheidung der Prinzipien.* This chapter, which I did not mention in my paper, comes to the conclusion that in fact even the threefold partition of the unproved mathematical principles in Euclid can be explained in the framework of Eleatic dialectic.

By the way, Professor Bernays' remark: 'As to the problem of the difference between *axioms* and the *common notions*, perhaps *axioms* are problematic starting points, and the *common notions* unproblematic starting points, concerning the general laws of *quantities*' seems certainly to be mistaken. The *κοιναὶ ἔννοιαι* (common notions) and *ἀξιώματα* (axioms) are two different names for the same group of principles. The Euclidean text at our disposal uses the term *κοιναὶ ἔννοιαι instead* of *ἀξιώματα.* (I discussed this problem at some length in my paper mentioned in the previous footnote.)

(6) Professor Kneale called attention to an interesting point: '. . . about order of priority between the use of *reductio ad absurdum* in metaphysics and its use in mathematics . . . we should not assume that mathematicians cannot use a logically valid pattern of reasoning in their work until some philosopher had written about it and told them that it is valid.' I think everybody will agree with this comment *in principle.* Theoretical considerations alone can scarcely decide whether philosophers or mathematicians first used indirect proof. But why do I then say that Greek mathematics took the method of indirect proof from Eleatic philosophy?

In Eleatic philosophy indirect proof has a very special central

[1] 'Anfänge des Euklidischen Axiomensystems', *Archives for History of Exact Sciences* I, pp. 37–106.

role. It would be a mistake to assume that Eleatic indirect proof appears first only with Zeno. It is true, as Aristotle says, that Zeno was the first dialectician; and since the core of Zeno's dialectic is indirect proof we could easily come to the conclusion that the first philosopher who consistently and skilfully used indirect proof was Zeno. But in fact this would be a misleading assertion. For indirect proof is already the core of Parmenidean philosophy. We should pay some attention also to the question: which were those assertions, theses, which the Eleatics proved by indirect methods? Did they first arrive at their philosophical theses by observation, experience or practice and only later prove the same theses indirectly? No! No observation, experience or practice can justify the theses of Eleatic philosophy. Just the opposite: observation, experience and practice would justify exactly the negation of the assertions of Eleatic philosophy. After all we always sense movement, change, generation and corruption, space and time. According to the Eleatics, however, there are no such things as movement, change, generation and corruption, space and time. All these things are only misleading illusions of the senses. But then all these Eleatic assertions can be proved only by indirect proofs and nothing else. Parmenides and the Eleatics show contradictions in the concepts they criticize. Their main argument is that what contradicts itself cannot be true and therefore they have to turn against so-called sober experience and, on the basis of theoretical reasoning, qualify the whole experienced natural world as mere appearance (*doxa*). *This is why indirect proof acquires central importance in Eleatic philosophy.* Of course we shall never be able to say who was the first person to justify an assertion indirectly. But if we ask which was the philosophical school that made indirect proof a method of central importance, and which was the philosophical school which anticipated this method, then there is no doubt that these were the Eleatics. And now since the whole terminology and method of indirect proof in mathematics is exactly identical with the terminology and method of indirect proof in Eleatic dialectic we can hardly resist the argument that the development of deductive mathematics has to be accounted for in terms of the influence of Eleatic philosophy.

I should only mention in this connection the following fact. I have already said that in Eleatic philosophy indirect proof was used for the justification of theses which one cannot possibly justify by observation and experience, but which rather are refuted by observation and experience. Therefore it is very interesting that in mathematics the oldest example of indirect proof applies similarly to a theoretical fact which one could never justify from experience. It is well known that the Pythagoreans showed by indirect proof that the side and the diagonal of the square are linearly incommensurable. But is there anything like linear incommensurability for sense experience, or for thought which only relies on sense experience? After all, in practice any two straight lines are commensurable: all that we need is to find a unit so small that when we use this unit we can no longer show the incommensurability of the two quantities. So the indirect proof of the Pythagoreans in this case proved a theorem which they could not have proved by any other method, just as the Eleatics could not have proved by any other method their philosophical theses.

(7) In connection with the problem of indirect proof Professor Kalmár was certainly right in stressing that 'it would be quite unhistorical to require of the ancient Greeks the fine distinctions [between kinds of indirect proof] which are made today in mathematical logic.' However I think that Professor Robinson answered his other question, namely: 'whether all 'indirect proofs' of the ancient Greeks were of the second kind [which does not require the principle of the excluded third] or whether they used also indirect reasonings which require the principle of the excluded third', when he reminded us that 'when Euclid shows that every composite number has a prime divisor [*Eucl. Elem.* VII, 31], he uses the method of infinite descent, concluding *'which is impossible for numbers'* . . . A contemporary intuitionist would [reject this] proof, as it stands, since it tries to establish the proof of a positive, existential statement.' To his further question, 'whether the Greeks were aware of this distinction' between the two kinds of indirect proof, I am afraid I cannot give any answer at the moment.

(8) I now come to the comment of Dr. Lucas who stressed Plato's role in the emergence of the axiomatic method: 'The

Eleatics certainly produced the argument by *reductio ad absurdum*: but it seems to me that it is Plato who put forward the ideal of axiomatization as something to be pursued fairly consciously.' I should like to mention that not very long ago historians of mathematics attributed a decisive role to Plato in the development of Greek mathematics. The title of a paper by H. G. Zeuthen published in 1913 is very characteristic: 'Sur les connaissances géometriques des Grecs *avant la réforme platonicienne*'. As the title shows, Zeuthen thought that Plato was a milestone in the history of Greek mathematics and that it was in fact Plato who started axiomatic deductive Greek mathematics. Although this view has been pushed more and more into the background in the last fifty years, even in 1963 a controversial and stimulating book by A. Frajese was published in Rome with the title *Platone e la matematica nel mondo antico* which puts forward roughly similar views to those of Mr. Lucas. I myself think that the axiomatization and deductivization of Greek mathematics is a process which was *independent* of Plato. I think that there is no evidence that Plato had any influence on this process. The relation between Plato and the circle of ideas of Euclidean mathematics has, I think, a very different character: for Plato mathematical thinking was a paradigm which had already been elaborated; on the other hand Plato is a successor to Eleatic philosophy just as Euclidean deductive mathematics is. In this respect I should like to refer again to the fifth chapter of my paper mentioned in the last footnote.

Unfortunately I cannot possibly go into a more elaborate characterization of pre-Platonic Greek dialectic in my present reply. I can only refer again to three of my earlier papers, 'Zur Geschichte der griechischen Dialektik', *Acta Antiqua Academiae Scientiarum Hungaricae*, Budapest I, pp. 377–410; 'Zur Geschichte der Dialektik des Denkens', *ibid.* II, pp. 17–62; and 'Zum Verständnis der Eleaten', *ibid.* II, pp. 243–289.

With this I come to the end of those questions where I could give a fairly clear-cut answer. From among those questions to which we can expect an answer only from future research I should like to mention two.

(a) The first has already been mentioned in passing. Professor Robinson asked whether the Greeks were aware of the distinction between the two kinds of indirect proof, one which does not require the principle of the excluded third, and the other which does. I am afraid that we need a much more thorough study of Greek mathematics and logic in order to have a clear picture of this problem.

(b) The other problem mentioned by Professor Bernays is perhaps still more interesting: 'We now know that there is no essential distinction between the things we prove and the things we assume without proof; for we can choose different sets of axioms. Did the Greeks know this? We should not simply assume that they did not: they were probably cleverer than we might think.' Unfortunately this is again a question where I do not know the answer, but perhaps I can say at least this: a few years ago I came across a paper by K. v. Fritz, 'Die APXAI in der griechischen Mathematik', *Archiv für Begriffsgeschichte*, Bd. 1, Bonn, 1955. In connection with this I studied those Aristotelian *loci* on the basis of which Fritz presents Aristotle's teaching about the so-called 'demonstrative sciences'. The quotations which he adduces from Aristotle certainly show that Aristotle *did not know* that mathematics chooses its unproved starting principles *arbitrarily*. It looks as if Aristotle was of the opinion that there are certain 'natural' simplest assertions which we cannot prove any further but the truth of which cannot possibly be doubted: mathematics has to choose these assertions as starting-points. However I agree with Professor Bernays that we should not assume that Aristotle's views were necessarily shared by Greek mathematicians. Future research has to investigate this historical problem much more thoroughly.

THE METAPHYSICS OF THE CALCULUS

ABRAHAM ROBINSON

University of California, Los Angeles

1. From the end of the seventeenth century until the middle of the nineteenth, the foundations of the Differential and Integral Calculus were a matter of controversy. While most students of Mathematics are aware of this fact they tend to regard the discussions which raged during that period entirely as arguments over technical details, proceeding from logically vague (Newton) or untenable (Leibniz) ideas to the methods of Cauchy and Weierstrass which meet modern standards of rigor. However, a closer study of the history of the subject reveals that those who actually took part in this dialogue were motivated or influenced quite frequently by basic philosophical attitudes. To them the problem of the foundations of the Calculus was largely a philosophical question, just as the problem of the foundations of Set Theory is regarded in our time as philosophical no less than technical. Thus, d'Alembert states in a passage from which I have taken the title of this address ([2]):

'La théorie des limites est la base de la vraie Métaphysique du calcul différentiel.'

It will be my purpose today to describe and analyse the interplay of philosophical and technical ideas during several significant phases in the development of the Calculus. I shall carry out this task against the background of Non-standard Analysis as a viable Calculus of Infinitesimals. This will enable me to give a more precise assessment of certain historical theories than has been possible hitherto.

The basic ideas of Non-standard Analysis are sketched in the next two sections. A comprehensive development of that theory will be found in [10]. The last chapter of that reference also contains a more detailed discussion of the historical issues raised in the present talk.

2. Let R be the field of real numbers. We introduce a formal language L in order to express within it statements about R. The precise scope of the language depends on the purpose in hand. We shall suppose here that we have chosen L as a very rich language. Thus, L shall include symbols for all individual real numbers, for all sets of real numbers, for all binary, ternary, quaternary, etc., relations between real numbers, and also for all sets and relations of higher order, e.g. the set of all binary relations between real numbers. In addition L shall include the connectives of negation, disjunction, conjunction and implication and also variables and quantifiers. Quantification will be permitted at all levels, but we may suppose, for the sake of familiarity, that type restrictions have been imposed in the usual way. Thus, L is the language of a 'type theory of order ω'. Within it one can express all facts of Real (or of Complex) Analysis. There is no need to introduce function symbols explicitly for to every function of n variables $y=f(x_1, \ldots, x_n)$ there corresponds an $n+1$-ary relation $F(x_1, \ldots, x_n, y)$ which holds if and only if $y=f(x_1, \ldots, x_n)$.

Let K be the set of all sentences in L which hold (are true) in the field of real numbers, R. It follows from standard results of Predicate Logic that there exists a proper extension $^*R$ of R which is a model of K, i.e. such that all sentences of K are true also in $^*R$. However, the statement just made is correct only if the sentences of K are interpreted in $^*R$ 'in Henkin's sense'. That is to say, when interpreting phrases such as 'for all relations' (of a certain type, universal quantification) or 'for some relation' (of a certain type, existential quantification) we take into account not the totality of all relations (or sets) of the given type but only a subset of these, the so-called *internal* or *admissible* relations (or sets). In particular, if S is a set or relation in R then there is a corresponding internal set or relation $^*S$ in $^*R$, where S and $^*S$ are denoted by the same symbol in L. However not all internal entities of $^*R$ are of this kind.

The *Non-standard model of Analysis* $^*R$ is by no means unique. However, once it has been chosen, the totality of its internal entities is given with it. Thus, corresponding to the set of natural numbers N in R, there is an internal set $^*N$ in $^*R$ such that $^*N$

is a proper extension of N. And $^*N$ has 'the same' properties as N, i.e. it satisfies the same sentences of L just as $^*R$ has 'the same' properties as R. N is said to be a *Non-standard model of Arithmetic.* From now on all elements (individuals) of $^*R$ will be regarded as 'real numbers', while the particular elements of R will be said to be *standard.*

$^*R$ is a non-archimedean ordered field. Thus $^*R$ contains non-trivial infinitely small (*infinitesimal*) numbers, i.e. numbers $a \neq 0$ such that $|a| < r$ for all standard positive r. (0 is counted as infinitesimal, trivially.) A number is *finite* if $|a| < r$ for some standard r, otherwise a is *infinite.* The elements of $^*N - N$ are the *infinite natural numbers.* An infinite number is greater than any finite number. If a is any finite real number then there exists a uniquely determined standard real number r, called the *standard part* of a such that $r - a$ is infinitesimal or, as we shall say also, such that r is infinitely close to a, write $r \simeq a$.

3. Let $f(x)$ be an ordinary ('standard') real-valued function of a real variable, defined for $a < x < b$, where a, b are standard real numbers, $a < b$. As we pass from R to $^*R$, $f(x)$ is extended automatically so as to be defined for all x in the open interval (a, b) *in* $^*R$. As customary in Analysis, we shall denote the extended function also by $f(x)$, but we may refer to it, by way of distinction, as '$f(x)$ in $^*R$', as opposed to the original '$f(x)$ in R'.

The properties of $f(x)$ in $^*R$ are closely linked to the properties of $f(x)$ in R by the fact that R and $^*R$ satisfy the same set of sentences, K. A single but relevant example of this interconnection is as follows.

Let $f(x)$ be defined in R, as above, and let x_0 be a standard number such that $a < x_0 < b$. Suppose that $f(x_0 + \xi) \simeq f(x_0)$, i.e. that $f(x_0 + \xi) - f(x_0)$ is infinitesimal, for all infinitesimal ξ, where $f(x)$ is now considered in $^*R$. Then we claim that for every standard $\varepsilon > 0$ there exists a standard $\delta > 0$ such that $|f(x_0 + \xi) - f(x_0)| < \varepsilon$ for all ξ such that $|\xi| < \delta$. – Indeed if ε is any standard positive real number then the statement,

'There exists an $\eta > 0$ such that for all ξ, $|\xi| < \eta$ implies $|f(x_0 + \xi) - f(x_0)| < \varepsilon$', can be formulated as a sentence X within L. Thus, either X or not-X holds in R. But if not-X held in R then it would belong to K and hence, would hold also in $^*R$. Since X holds in $^*R$, by its definition, we conclude that actually X holds also in R. And any f which realizes η in R must be standard since there are no other numbers in R. This proves our assertion.

We may also prove the converse, i.e. if for every standard $\varepsilon > 0$ there exists a standard $\delta > 0$ such that $|f(x_0+\xi)-f(x_0)| < \varepsilon$ for all ξ such that $|\xi| < \delta$ in R, then $f(x_0+\xi) \simeq f(x_0)$ for all infinitesimal ξ in $^*R$. *This shows that $f(x)$ is continuous at x_0 in R if and only if $f(x_0+\xi)$ is infinitely close to $f(x_0)$ in $^*R$.*

Similarly, it can be shown that $f(x)$ is differentiable at x_0 if and only if the ratios $(f(x_0+\xi)-f(x_0))/\xi$ have the same standard part, d, for all infinitesimal $\xi \neq 0$, and d is then the derivative of $f(x)$ at x_0 in the ordinary sense.

For a last example, let $\{s_n\}$ be an infinite sequence of real numbers in R. On passing from R to $^*R$, $\{s_n\}$ is extended so as to be defined also for infinite natural numbers n. Let s be a standard real number. It can then be proved that s is the limit of $\{s_n\}$ in the ordinary sense, $\lim_{n\to\infty} s_n = s$ if and only if s_n is infinitely close to s (or, which is the same, if s is the standard part of s_n) for all infinite natural numbers n.

The above examples may suffice in order to give a hint how the Differential and Integral Calculus can be developed within the framework of Non-standard Analysis.

4. It appears that Newton's views concerning the foundations of the Calculus were somewhat ambiguous. He referred sometimes to infinitesimals, sometimes to moments, sometimes to limits and sometimes, and perhaps preferentially, to physical notions. But although he and his successors remained vague on the cardinal points of the subject, he did envisage the notion of the limit which, ultimately, became the cornerstone of Analysis. By contrast, Leibniz and his successors wished to base the Calculus, clearly and unambiguously, on a system which includes infinitely small quantities. This approach is crystallized in the first sentence of the '*Analyse des infiniment petits pour l'intelligence des lignes courbes*' by the Marquis de l'Hospital. We mention in passing that de l'Hospital, who was a pupil of Leibniz and John Bernoulli, acknowledged his indebtedness to his two great teachers.

De l'Hospital's begins with a number of definitions and axioms. We quote (translated from [7]):

'Definition I. A quantity is *variable* if it increases or decreases continuously; and, on the contrary, a quantity is *constant* if it remains the same while other quantities change. Thus, for a parabola, the ordinates and abscissae are variable quantities while the parameter is a constant quantity.'

'Definition II. The infinitely small portion by which a variable increases or decreases continuously is called its difference . . .'

For *difference* read *differential*. There follows an example with reference to a diagram and a corollary in which it is stated as evident that the differential of a constant quantity is zero. Next, de l'Hospital introduces the differential notation and then goes on –

'First requirement or supposition. One requires that one may substitute for one another [*prendre indifféremment l'une pour l'autre*] two quantities which differ only by an infinitely small quantity: or (which is the same) that a quantity which is increased or decreased only by a quantity which is infinitely smaller than itself may be considered to have remained the same . . .'

'Second requirement or supposition. One requires that a curve may be regarded as the totality of an infinity of straight segments, each infinitely small: or (which is the same) as a polygon with an infinite number of sides which determine by the angle at which they meet, the curvature of the curve . . .'

Here again we have omitted references to a diagram.

In order to appreciate the significance of these lines we have to remember that, when they were written, mathematical axioms still were regarded, in the tradition of Euclid and Archimedes, as empirical facts from which other empirical facts could be obtained by deductive procedures; while the definitions were intended to endow the terms used in the theory with an empirical meaning. Thus (contrary to what a scheme of this kind would signify in our time) de l'Hospital's formulation implies a belief in the reality of the infinitely small quantities with which it is concerned. And the same conclusion can be drawn from the preface to the book –

'Ordinary Analysis deals only with finite quantities: this one [i.e. the Analysis of the present work] penetrates as far as infinity itself. It compares the infinitely small differences of finite quantities; it discovers the relations between these differences; and in this way makes known the relations between finite quantities, which are, as it were, infinite compared with the infinitely small quantities. One may even say that this Analysis extends beyond infinity: For it does not confine itself to the infinitely small differences but discovers the relations between the differences of these differences, . . .'

It is this robust belief in the reality of infinitely small quantities which held sway on the continent of Europe through most of the eighteenth century. And it is this point of view which is commonly believed to have been that of Leibniz. However, although Leibniz was indeed responsible for the technique and notation of this Calculus of Infinitesimals his ideas on the foundations of the subject were quite different and considerably more subtle. In fact, we know from Leibniz' correspondence that he was critical of de l'Hospital's belief in the reality of infinitesimals and even more critical of Fontenelle's emphatic affirmation of this opinion.

Leibniz' own view, as published in 1689 [8] and as repeated and elaborated subsequently in a number of letters, may be summarized as follows. While approving of the introduction of infinitely small and infinitely large quantities, Leibniz did not consider them as real, like the ordinary 'real' numbers, but thought of them as ideal or fictitious, rather like the imaginary numbers. However, by virtue of a general principle of continuity, these ideal numbers were supposed to be governed by the same laws as the ordinary numbers. Moreover, Leibniz maintained that his procedure differed from 'the style of Archimedes' only in its language [*dans les expressions*]. And in describing 'the style of Archimedes' i.e. the Greek method of exhaustion, he used the following perfectly appropriate, yet strikingly modern, phrase (translated from [9]):

'One takes quantities which are as large or as small as is necessary in order that the error be smaller than a given error [*pour que l'erreur soit moindre que l'erreur donnée*] . . .'

However, Leibniz, like de l'Hospital after him, stated that two quantities may be accounted equal if they differ only by an amount which is infinitely small relative to them. And on the other hand, although he did not state this explicitly within his axiomatic framework, de l'Hospital, like Leibniz, assumed that the arithmetical laws which hold for finite quantities are equally valid for infinitesimals. It is evident, and was evident at the time, that these two assumptions cannot be accommodated simultaneously within a consistent framework. They were widely accepted nevertheless, and maintained themselves for a considerable length of time since it was found that their judicious and selective use was so very fruitful. However, Non-standard Analysis shows how a relatively slight modification of these ideas leads to a consistent theory or, at least, to a theory which is consistent relative to classical Mathematics. Thus, instead of claiming that two quantities which differ only by an infinitesimal amount, e.g. x and $x+\mathrm{d}x$, are actually equal, we find only that they are equivalent in a well-defined sense, $x+\mathrm{d}x \simeq x$ and can thus be substituted for one another in some relations but not in others. At the same time, the assertion that finite and infinitary quantities have 'the same' properties is explicated by the statement that both R and $^*R$ satisfy the set of sentences K. And if we ask, for example, whether $^*R$ (like R) satisfies Archimedes' axiom then the answer depends on our interpretation of the question. If by Archimedes' axiom we mean the statement that from every positive number a we can obtain a number greater than 1 by repeated addition –

$$a+a+\ldots+a \ (n \text{ times}) > 1,$$

where n is an ordinary natural number, then $^*R$ does not satisfy the axiom. But if we mean by it that for any $a>0$ there exists a natural number n (which may be infinite) and that $n \cdot a > 1$, then Archimedes' axiom does hold in $^*R$.

5. In the view of many, including the author, the problem of the nature of infinitary notions is still of central importance in the Philosophy of Mathematics. To a logical positivist, the entire argument over the reality of a mathematical structure may seem pointless but even he will have to acknowledge the historical importance of the issue. To de l'Hospital, the infinitely small and large quantities (which were still thought of as geometrical entities) represented the actual infinite. On the other hand, Leibniz stated specifically that although he believed in the actual infinite in other spheres of Philosophy, he did not assume its existence in Mathematics. He also said that he accepted the potential (or as he put it, referring to the schoolmen, 'syncategorematic') infinite as exemplified, in his view, in the number of terms of an infinite series. To sum up, Leibniz accepted the ideal, or fictitious, infinite; accepted the potential infinite; and within Mathematics, rejected or at least dispensed with, the actual infinite.

Like the proponents of the new theory, its critics also were motivated by a combination of technical and philosophical considerations. Berkeley's 'Analyst' ([3]; compare [11]) constitutes a brilliant attack on the logical inadequacies both of the Newtonian Theory of Fluxions and of the Leibnizian Differential Calculus. In discrediting these theories, Berkeley wished to discredit also the views of the scientists on theological matters. But beyond that, and more to the point, Berkeley's distaste for the Calculus was related to the fact that he had no place for the infinitesimals in a philosophy dominated by perception.

6. The second half of the eighteenth century saw several attempts to put the Calculus on a firm footing. However, apart from d'Alembert's affirmation of the importance of the limit concept (and, possibly, some of L. N. M. Carnot's ideas, which may have influenced Cauchy), none of these made a contribution of lasting value. Lagrange's attempt to base the entire subject on the Taylor series expansion was doomed to failure although, indirectly, it may have had a positive influence on the development of the idea of a formal power series.

It is generally believed that it was Cauchy who finally put the Calculus on rigorous foundations. And it may therefore come as a surprise to learn that infinitesimals still played a vital role in his system. I quote from Cauchy's *Cours d'Analyse* (translated from [6]):

'When speaking of the continuity of functions, I was obliged to discuss the principal properties of the infinitesimal quantities, properties which constitute the foundation of the infinitesimal calculus . . .'

However, Cauchy did not regard these entities as basic but tried to derive them from the notion of a variable:

'A variable is a quantity which is thought to receive successively different values . . .'

'When the successive numerical values of a variable decrease indefinitely so as to become smaller than any given number, this variable becomes what is called an *infinitesimal* [*infiniment petit*] or an infinitely small quantity.'

At the same time the *limit* of a variable (when it exists) is defined as a fixed value which is approached by the variable so as to differ from it finally as little as one pleases. It follows that a variable which becomes infinitesimal has zero as limit.

Cauchy did not wish to regard the infinitesimals as numbers. And the assumption that they satisfy the same laws as the ordinary numbers, which had been stated explicitly by Leibniz, was rejected by Cauchy as unwarranted. Moreover, Cauchy stated, on a later occasion, that while infinitesimals might legitimately be used in an argument they had no place in the final conclusion.

However, Cauchy's professed opinions in these matters notwithstanding, he did in fact treat infinitesimals habitually as if they were ordinary numbers and satisfied the familiar rules of Arithmetic. And, as it happens, this procedure led him to the correct result in most cases although there is a famous and much discussed situation in the theory of series of functions in which he was led to the wrong conclusion. Here again, Non-standard Analysis, in spite of its different background, provides a remarkably appropriate tool for the discussion of Cauchy's successes and failures.

For example, the fact that a function $f(x)$ is continuous at a point x_0 if the difference $f(x_0+\xi)-f(x_0)$ is infinitesimal for infinitesimal ξ, which is a *theorem* of Non-standard Analysis (see section 3 above), is also a precise explication of Cauchy's notion of continuity. On the other hand, in arriving at the wrong conclusion that the sum of a series of continuous functions is continuous provided it exists, Cauchy used the unwarranted argument that if

$\lim_{n\to\infty} s_n(x) = s(x)$ over an interval then $s_n(x_0) - s(x_0)$ is, *for all* x_0 *in the interval*, infinitesimal for infinite n. In Non-standard Analysis, it turns out that this is true for standard (ordinary) $s_n(x)$, $s(x)$ and x_0, but not in general for non-standard x_0, e.g. not if $x_0 = x_1 + \xi$ where x_1 is standard and ξ is infinitesimal.

In order to appreciate to what extent Cauchy regarded the infinitesimals as an integral part of his system, it is instructive to consider his definition of a derivative. To him, $f'(x)$, wherever it exists, is the limit of the ratio

$$\frac{\Delta y}{\Delta x} = \frac{f(x+\xi) - f(x)}{\xi}$$

where ξ *is infinitesimal*. In the standard modern approach the assumption that ξ is infinitesimal is completely redundant or, more precisely, meaningless. The fact that it was nevertheless introduced explicitly by Cauchy shows that his mental image of the situation was fundamentally different from ours. Thus, it would appear that, to his mind, a variable does not attain the limit zero directly but only after travelling through a region of infinitesimals.

We have to add that in our 'classical' framework the entire notion of a variable in Cauchy's sense, as a mathematical entity *sui generis*, has no place. We might describe a variable, in a jocular mood, as a function which has lost its argument, while Cauchy's infinitesimals still are, to use Berkeley's famous phrase, the ghosts of departed quantities. But such carping criticism does not help us to understand the just recognition accorded to Cauchy's achievement, which is still thought by many to have resolved the fundamental difficulties that had beset the Calculus previously.

If we wish to find the reasons for Cauchy's success we have to consider, once again, both the technical-mathematical and the basic philosophical aspects of the situation. Cauchy established the central position of the limit concept for good. It is true that d'Alembert, who had emphasized the importance of this concept some decades earlier, in a sense went further than Cauchy by stating (translated from [1]):

'We say that in the *Differential* Calculus there are no infinitely small quantities at all . . .'

But apparently d'Alembert did not work out the consequences of his general principles; while the vast scope and the subtlety of Cauchy's mathematical achievement showed to the world that his tools enabled him to go farther and deeper than his predecessors. He introduced these tools at a time when the great achievements of the earlier and technically more primitive method of infinitesimals had become commonplace. Thus, the momentum which had enabled that method to disregard earlier attacks such as Berkeley's was exhausted before the end of the eighteenth century and due attention was again given to its logical weaknesses (which had been there, for all to see, all the time). These weaknesses had been associated throughout with the introduction of entities which were commonly regarded as denizens of the world of actual infinity. It now appeared that Cauchy was able to remove them from that domain and to base Analysis on the potential infinite (compare [4] and [5]). He did this by choosing as basic the notion of a variable which, intuitively, suggests potentiality rather than actuality. And so it happened that a grateful public was willing to overlook the fact that, from a strictly logical point of view, the new method shared some of the weaknesses of its predecessors and, indeed, introduced new weaknesses of its own.

7. When Weierstrass (who had been anticipated to some extent by Bolzano) introduced the δ,ε-method about the middle of the nineteenth century he maintained the limit concept in its central place. At the same time, Weierstrass' approach is perhaps closer than Cauchy's to the Greek method of exhaustion or at least to the feature of that method which was described by Leibniz ('*pour que l'erreur soit moindre que l'erreur donnée*', see section 4 above). On the issue of the actual infinite versus the potential infinite, the δ,ε-method did not, as such, force its proponents into a definite position. To us, who are trained in the set-theoretic tradition, a phrase such as 'for every positive ε, there exists a positive δ . . .' does in fact seem to contain a clear reference to a well-defined infinite totality, i.e. the totality of positive real numbers. On the other hand, already Kronecker made it clear, in his lectures, that to him the phrase meant that one could *compute* for, every *specified* positive ε, a positive δ with the required property. However, it was not then known that the abstract and the constructive approaches actually lead to different theories of Analysis, so that a mathematician's inability to provide a procedure for computing a function whose existence he has proved by abstract arguments is not necessarily due to his personal inadequacy.

At the same time it is rather natural that Set Theory should have arisen, as it did, from the consideration of certain problems of Analysis which required the further clarification of basic concepts. And its creator, Georg Cantor, argued forcefully and in great detail that Set Theory deals with the actual infinite. Nevertheless, Cantor's attitude towards the theory of infinitely small quantities was entirely negative, in fact he went so far as to claim that he could disprove their existence by means of Set Theory. I quote (translated from [4]):

'*The fact of* [the existence of] *actually-infinitely large numbers is not a reason for the existence of actually-infinitely small quantities; on the contrary, the impossibility of the latter can be proved precisely by means of the former.*

'Nor do I think that this result can be obtained in any other way *fully* and *rigorously*.'

The misguided attempt which is summed up in this quotation was concerned not only with the past but was directed against P. du Bois-Reymond and O. Stolz who had just re-established a modest but rigorous theory of non-Archimedean systems. It may be recalled that, at the time, Cantor was fighting hard in order to obtain recognition for his own theory.

Cantor's belief in the actual existence of the infinities of Set Theory still predominates in the mathematical world today. His basic philosophy may be likened to that of de l'Hospital and Fontenelle although their infinite quantities were thought to be concrete and geometrical while Cantor's infinities are abstract and divorced from the physical world. Similarly, the intuitionists and other constructivists of our time may be regarded as the heirs to the Aristotelian traditions of basing Mathematics on the potential infinite. Finally, Leibniz' approach is akin to Hilbert's original formalism, for Leibniz, like Hilbert, regarded infinitary entities as ideal, or fictitious, additions to concrete Mathematics. Thus, we may conclude this talk with the observation that although the very subject matter of foundational research has changed radically over the last two hundred years, there is a remarkable permanency in the concern with the infinite in Mathematics and in the various philosophical attitudes which have been adopted towards this notion.

References

[1] J. Le R. d'Alembert, article 'Différentiel' in *Encyclopédie méthodique ou par ordre de matières* (Mathématiques) 3 vols., Paris–Liège 1784–1789.

[2] J. Le R. d'Alembert, article 'Limite' in *Encyclopédie méthodique ou par ordre de matières* (Mathématiques) 3 vols., Paris–Liège 1784–1789.

[3] G. Berkeley, *The Analyst*, 1734, Collected works, vol. 4 (ed. A. A. Luce and T. E. Jessop) London 1951.

[4] G. Cantor, *Mitteilungen zur Lehre vom Transfiniten*, 1887–1888, Gesammelte Abhandlungen ed. E. Zermelo, Berlin 1932, pp. 378–439.

[5] E. Carruccio, I Fondamenti dell'Analisi matematica nel pensiero di Agostino Cauchy, *Bolletino dell'Unione Matematica Italiana* ser. 3, vol. 12, 1957, pp. 298–307.

[6] A. Cauchy, *Cours d'Analyse de l'Ecole Royale Polytechnique*, 1re partie, Analyse Algébrique, 1821 (Oeuvres complètes ser. 2, vol. 3).

[7] G. F. A. de l'Hospital, *Analyse des infiniment petites pour l'intelligence des lignes courbes*, Paris (1st ed. 1696) 2nd ed. 1715.

[8] G. W. Leibniz, Tentamen de motuum coelestium causis, *Acta Eruditorum*, 1689, Mathematische Schriften (ed. C. I. Gerhardt) vol. 5, 1858, pp. 320–328.

[9] G. W. Leibniz, Mémoire de M. G. G. Leibniz touchant son sentiment sur le calcul différentiel, *Journal de Trévoux*, 1701, Mathematische Schriften (ed. C. I. Gerhardt) vol. 5, 1858, p. 350.

[10] A. Robinson, *Non-standard Analysis*, Studies in Logic and the Foundations of Mathematics, Amsterdam, 1966.

[11] E. W. Strong, Mathematical reasoning and its objects, *George Berkeley lectures*, University of California Publications in Philosophy, vol. 29, Berkeley and Los Angeles, 1957, pp. 65–88.

DISCUSSION

Peter Geach: *Infinity in scholastic philosophy.*

Leibniz, as Robinson's quotation shows, assimilated the distinction of actual and potential infinity to the distinction of categorematic and syncategorematic infinity. This was common form in the scholasticism of his own day (it is to be found already in Suarez [1]); and yet by the standards of an older scholasticism it was a confusion so gross as might excuse an enemy of scholasticism for echoing Lord Chesterfield's remark to one of the College of Heralds, that the foolish man did not even understand his own foolish business.

The distinctions are not even the same *sort* of distinction. The distinction between actual and potential infinity is a distinction between two ways in which outside things, *res extra*, could be said to be infinite. 'Categorematic' and 'syncategorematic' on the other hand are words used to describe (uses of) words in a language; an infinite multitude, say, can no more be syncategorematic than it can be pronominal or adverbial. To be sure, the confusion is explicable. A *categorematic* use of a word is a use of it so that it can be understood as a logical subject or predicate; and just those things are *actually* infinite of which the word 'infinite' taken *categorematically* can with truth be predicated *sans phrase* (*simpliciter*). But this does not make the confusion excusable – especially as there is no such close connexion between the potentially infinite and the syncategorematic use of 'infinite'.

I shall give an example of a sentence calling for the categorematic-syncategorematic distinction and then go on to the medieval rationale of the distinction. When we read in Spinoza that there are infinite Divine attributes, we need to know whether he meant that each attribute is an infinite attribute, or, that there are infinitely many attributes; in fact the latter was his meaning. A medieval scholastic would have said that 'infinite' is taken categore-

[1] Suarez, *De Incarnatione*, disp. 26, s. 4, n. 5.

matically in the first case and syncategorematically in the second case. The example may serve to stop us from thinking in terms of actuality and potentiality; for of course nothing was further from Spinoza's mind than that the infinity of Divine attributes was only potential; all the same, if 'infinite' in 'There are infinite Divine attributes' means 'infinitely many', there is no choice but to parse it as a syncategorematic use of 'infinite'.

Syncategorematic words, for medieval logicians, were words that give form to propositions, such as the copula, negation, quantifiers, and connectives [1]. (Later scholastics also count e.g. adverbs like 'badly' and possessives like 'Cicero's' as syncategorematic; this extended application seems to me misleading.) 'Infinite' in the syncategorematic sense is explicitly assimilated to the quantifiers; and rightly so – 'there are infinitely many' is plainly an expression of the same semantical category as 'there are some' or 'there are none' [2].

The closest connexion that can be made between the syncategorematic use of 'infinite' and the potentially infinite is this: the scholastic account of the syncategorematic word 'infinite' makes it licit to use it without (at least obviously) introducing actually infinite numbers. For 'there are infinite –s' was expounded as meaning 'there are not so many –s that there are no more' (*non sunt tot quin sint plura*); and in modern quantificational terms this comes out as 'for no n are there no more than n –s', where the range of the variable 'n' is restricted to *finite* cardinals.

HANS FREUDENTHAL: *Technique versus metaphysics in the calculus.*

There is more metaphysics in Leibniz' speculations on Calculus than is usually known, e.g. attempts to understand the relation of body and soul by an analogy with that between a magnitude

---

[1] J. Reginald O'Donnell C.S.B. 'The Syncategoremata of William of Sherwood', *Mediaeval Studies* III, 1941, pp. 48, 54 f.

[2] Walter Burleigh, *De puritate artis logicae*, Tractatus brevior. See also the edition of the *Tractatus longior* by P. Bochner, O.S.B., St. Bonaventure, N.Y., 1955: pp. 259f.

and its differential. The genetic theory of preformation which asserted that the new creature has been preformed in its progenitors and particularly the whole of mankind in Adam, led to the idea of the differential of a genus, from which the genus developed as its integral.

Differentials were well-known in antiquity (atomic lines). Archimedes used them as a heuristic tool. In modern times they were reintroduced by Kepler, Cavalieri (indivisibles), and others, and systematically used by Newton's and Leibniz' predecessors. Their methods were more or less geometrical. This is particularly true of Pascal who behaved idiosyncratically towards Cartesian methods. Leibniz' starting point was an integral transformation he found in Pascal's work and stripped of its geometrical clothing. The gist of Leibniz' efforts was the thorough algebraisation of calculus. The result was an easy and prolific formalism, more practical than Newton's, and rapidly accepted by most creative mathematicians.

A. HEYTING: *Technique versus metaphysics in the calculus.*

One of the main points of the lecture was, that questions which originally were considered as metaphysical can later on be considered as merely technical questions. I would like to relativize a little further the difference between metaphysical and technical questions, and perhaps also from another point of view to make it more absolute. To begin with the latter, it is clear that the things which we write on the blackboard, simply considered as signs, are purely technical; on the other side it is clear that the theological considerations by which Cantor motivated his notion of the actual infinite, were metaphysical in nature. But there is quite a gradual scale of notions between the purely technical and the almost purely metaphysical. As soon as we give any interpretation to the signs, we introduce metaphysical or at least philosophical notions. If you consider non-standard analysis as technical, at the same time you consider it as an interpreted set theory, and from that point of view it contains some metaphysics. I agree that what now are considered as questions of philosophical nature, for instance the opposition between constructivism and Cantorism or Platonism, can from a certain point of view be considered as technical, because all these lines of thought are expressed in developments which can be considered from a purely formal point of view. But on the other hand, what now are considered as technicalities can be considered later on as more philosophical; the philosophical implications can become more relevant in the future.

Y. Bar-Hillel: *The irrelevance of ontology to mathematics.*

I don't think we have to wait for the future in order to make up our minds on the ontological (or metaphysical, or philosophical) status of the main mathematical entities. In particular, I don't think we have to wait much longer before realizing that the current practice of many philosophers of mathematics who use such terms as 'real', 'ideal' or 'fictional' to qualify mathematical entities is of little help towards the clarification of the methodological issues involved (in complete analogy to the situation in the empirical sciences). In my view, this is just another instance of the confusions created by using the material mode of speech on an inappropriate occasion, and this seems to me to have been definitely the case in the Hilbert–Bernays way of talking about ideal mathematical entities. (I am aware of the history of this usage.) The sooner we stop being concerned with the 'ontological' problems of recent set theory, which are nothing but the product of an unhappy mode of speech, the sooner will we get down to discussing the real issues raised by recent developments.

M. Bunge: *Non-standard analysis and the conscience of the physicist.*

One century elapsed between the execution and burial of infinitesimals by the ε-δ revolution, and their resurrection in non-standard analysis. The historian may rejoice upon finding that some of the intuitions of Leibniz, Newton, Euler and their followers, though coarse, were after all not stupid. And the physicist may feel relieved. Indeed, he has never ceased to use infinitesimals, e.g. in setting up differential equations representing physical processes. But he has done it with a bad conscience ever since rumours of the Dedekind–Weierstrass revolution reached his ear. He can now refer to non-standard analysis for the rigorous justification of his intuitive infinitesimals, just as he refers to the theory of distributions for the legalization of the various delta 'functions' which his physical intuition led him to introduce. These two cases illustrate the thesis that intuition is not to be rejected provided one can control it rationally. (See the commentator's *Intuition and Science*, 1962.) The only thing to be feared, in connection with the modern heir to the old infinitesimal, is that the standard calculus teacher may feel entitled to teach analysis the easy way (as it was still done at the turn of the century) by identifying modern infinitesimals with old infinitesimals and the latter with differentials. Happily the news of this Robinsonian resurrection will not spread that quickly.

ABRAHAM ROBINSON: *Reply*.

Commenting first on the philosophical points raised by Bar-Hillel and Heyting, it will be evident that in my paper I have dealt with questions of reality in Mathematics from the detached point of view of a historian. However, I am willing to go further and commit myself to the point of stating that in my view these problems should not be discussed in the cavalier fashion advocated by Bar-Hillel. As to what is technical and what is essential, I certainly did not want to suggest that the very differences of opinion between constructivists and platonists are merely technical. But it seems to me likely that questions of detail within transfinite set theory such as the correctness of the continuum hypothesis or the existence of very large cardinals, will come to be regarded as philosophically irrelevant, although I yield to no one in my admiration for the ingenious methods which have been devised to cope with these problems.

In reply to Bunge and Freudenthal, I wish to emphasize that

Leibniz' infinitesimals, like my own, are not in-divisible and in this respect should be distinguished from indivisibles. However, it is true that, historically, the distinction is blurred. I may add that Pascal (writing as 'Monsieur Dettonville') anticipated Leibniz in claiming that the method of the ancients and the method of indivisibles differed only in their expression (*manière de parler*).

I gather that Professor Geach accepts my suggestion that in Leibniz' mind syncategorematic infinity and potential infinity were the same. On my part, I have enjoyed his exposition of the original medieval point of view. However, as he himself points out in his closing remarks, some connection can be made between the two concepts so that their identification is at least not entirely fortuitous.

ON A FREGEAN DOGMA *

FRED SOMMERS

Brandeis University, Waltham, Mass.

1. In the following passage Russell states an accepted and familiar thesis:

> The first serious advance in real logic since the time of the Greeks was made independently by Peano and Frege—both mathematicians. Traditional logic regarded the two propositions 'Socrates is mortal' and 'All men are mortal' as being of the same form; Peano and Frege showed that they are utterly different in form. The philosophical importance of logic may be illustrated by the fact that this confusion—which is still committed by most writers—obscured not only the whole study of the forms of judgment and inference, but also the relation of things to their qualities, of concrete existence to abstract concepts, and to the world of Platonic ideas ... Peano and Frege, who pointed out the error did so for technical reasons ... but the philosophical importance of the advance which they made is impossible to exaggerate.

In what follows I wish to be understood as criticising the quantificational "translation" of general categoricals like 'All men are mortal' only insofar as this is represented as exhibiting such statements to have a different logical form from singular predications. I am not criticising quantification theory as an indispensable logical tool, especially for inference involving statements of more than one variable. The standard general categoricals however are not of this type; it is for example well-known that quantification is not needed for syllogistic inference. What is not known is that we can treat the categoricals as simple subject-predicate statements on an exact par with singular predications. There is therefore no good logical reason for saying that general and singular statements must differ in logical form.

The doctrine that (1) 'Socrates is mortal' and (2) 'Men are mortal'

---

* This paper is the result in part of research sponsored by the Air Force Office of Scientific Research, U.S.A.F. and the Office of Naval Research, US Navy under Grants AF–AFOSR 881–65 and AF–AFOSR 987–66 (NR 348–12).

differ in logical form assumes that the following is the corect account of what these statements say: (a) Both say that 'is mortal' is true of some, thing or things; the first says it is true of Socrates; the second that it is true of whatever 'is a man' is true. It follows (b) that the logical form of the second statement differs from that of the first. For while the first is a simple predication, the second is a "quantified" statement.

2. We note that the assumption (a) that (2) is *like* (1) in affirming *mortal* in the singular forces on us the doctrine (b) that (1) and (2) have different logical forms. For by this assumption we are debarred from construing (2) as saying that 'are mortal' is true of men – an interpretation that puts (2) on a logical par with (1) as a simple predication. On this older interpretation, any statement of the form *S*'s are *P*'s is about the *S*'s in exactly the same way as '*S* is a *P*' is about *S*. Both statements affirm *P* of their respective subjects and this affirmation as such is neither singular nor plural. Of course when the subject term is singular the grammatical predication is singular and when it is a plural term, the grammatical predication will be plural. But, logically, predication is neither. Contrast this with the standard (contemporary) doctrine that, any statement of the form '*S*'s are *P*'s' must be understood as "really" saying 'is a *P*' is true of each thing etc. The assumption is that all predication is *logically* singular. I shall call this assumption the dogma of singular predication. It is a dogma worth exposing and discarding.

3. (1) Mortal (Socrates) and (2) Mortal (Men) clearly have the same logical form. Neither statement is quantified. This is the "error" Russell speaks of; we are considering a general statement of the form *S*'s are *P*'s as being of the same form as a singular statement of the form *S* is *P*. However, let us persist in this "error" and proceed to dequantify all four standard general categorical statements. To do this successfully we shall have to interpret all four as simple predications with a plural subject. All four say something about the *S*'s. Any difference among *A*, *E*, *I* and *O* statements is due to what is being *said* about the *S*'s. It will therefore occur in the predicate only. As for the subject, it is worth

emphasising that the plural term does not denote an individual of any type. In saying that men are mortal we do not say anything about something called the class of men. Nor are we *saying* of each man that he is mortal. We readily grant that 'are mortal' is true of men only if 'is mortal' is true of each man. But this must not be taken as an admission that 'Men are mortal' "really says" of each man that he is mortal. We maintain a sharp distinction between what 'Men are mortal' "really says" and the conditions for its truth. It straightforwardly says of men that they are mortal; this is in fact true only if each man is mortal. But also it is true only if Socrates is mortal and few philosophers would wish to argue that Socrates being mortal is something expressed by 'Men are mortal' though it is certainly a condition for the truth of that general statement.

4. We turn now to the job of formulating a dequantified predicative version of the four categoricals. The *A*-proposition has already been dealt with. It has the form *Ps* with '*S*' as a plural or general term. Since the *O*-statement denies what the *A*-statement affirms we may readily dequantify it along with the *A*-statement. This gives us two of the four:

A. *Ps* . . . . . (All the) *S*'s *are* *P*'s
O. *P's* . . . . . (All the) *S*'s *aren't* *P*'s

Admittedly, '*S*'s aren't *P*'s' is ambiguous in ordinary discourse. But I am insisting here on a logically strict interpretation of this statement, according to which 'aren't *P*'s' is true of the *S*'s whenever 'are *P*'s' is false of them. Since 'are *P*'s' is false of the *S*'s in the case where some *S*'s are not *P*'s, the statement '*S*'s aren't *P*'s' is the denial of '*S*'s are *P*'s' [1]. Thus '*P's*' behaves like '$\sim Ps$', the formal contradictory of *Ps*. For many contexts we may use the sign of negation to express predicate denial. I wish, however, to point out that denying a predicate term of a subject is logically different from negating the statement affirming that term of the

[1] '(All) glittering things aren't gold' is the predicative contradictory of '(All) glittering things are gold'. Because of the ambiguity between 'are not-gold' and 'aren't gold', this form is generally avoided.

same subject. This difference between predicate denial and statement negation is extremely important and indeed fundamental for certain contexts. But I shall not insist on it now and those who read '*P's*' as the negation of '*Ps*' will not go astray. Turning now to the *E*-statement and its *I*-denial we confront a difficulty, that gives a good clue to one source of the dogma of singular predication. Let *E* be the statement 'No *S* are *P*'. Clearly we cannot interpret *E* as saying of no *S*'s that they are *P*. That would mean that 'are *P*' is true of no *S*'s and even if this makes some sort of sense it is not an interpretation of *E* in which something is being said of the *S*'s. Nor can we interpret *E* as saying of the *S*'s that they aren't *P*. In that case *E* would be claiming that 'aren't *P*' is true of the *S*'s but then *E* would not be distinguished from *O*. It seems then that we are forced to construe *E* as saying 'is not *P*' is true of *each* *S* (or each thing of which 'is an *S*' is true). But this means we have been forced into singular predication and a "quantified" interpretation of the *E*-statement. Let us however continue to insist on a predicative, "dequantified" interpretation of *E*, that is, one in which *E* is read as predicating something of the *S*'s. What does *E* say about the *S*'s? Clearly that 'are not-*P*' is true of them. The so-called obverse of 'No *S* are *P*' expresses this explicitly: '*S*'s are not-*P*'. We see then that the logically perspicuous contrary of the *A*-statement is precisely that statement which affirms the contrary of what the *A*-statement affirms. Now the *A*-statement affirms *P* of the *S*'s, the *E*-statement affirms *not-P* or *un-P* of them. One reason logicians have ignored the possibilities of using plural predication for a dequantified interpretation of the four categoricals is their refusal to accord logical recognition to contrariety as a distinct logical relation between terms. If one insists on treating '(*X* is un-*P*)' as if it were '$\sim$ (*X* is *P*)', the *E*-statement cannot be dequantified. Nor will then one be moved to challenge the doctrine of different logical forms or the underlying dogma of singular predication.

5. Since the *I*-statement is the denial of the *E*-statement, the dequantified, predicative schedule for the four standard general categoricals is now complete.

		Perspicuous expression	Vernacular expression
A.	Ps	*S*'s are *P*'s	All *S* are *P*.
E.	$\overline{P}s$	*S*'s are un-*P*'s	No *S* are *P*.
I.	$\overline{P}'s$	*S*'s aren't un-*P*'s	Some *S* are *P*.
O.	$P's$	*S*'s aren't *P*'s	Some *S* are not *P*.

All four categoricals are about (all) the *S*'s. In *A*, *P* is affirmed, in *E*, $\overline{P}$, its contrary, is affirmed. I shall therefore say that *E* *contraffirms* what *A* affirms. In *I*, the contrary of *P* is denied of the *S*'s. I shall say that *I contradenies A*. In *O*, *P* is denied. The two fundamental modes of predication are affirmation and denial. *A* and *E* are affirmations, *I* and *O* are denials. But since either a term or its contrary may be affirmed or denied we get four logically distinct ways of predicating a term. Quantification is eliminated or absorbed by these four predicative modes. They are: affirmation, contraffirmation, contradenial, and denial. Contrariety, it should be noted, is a *relation*. An affirmative statement, taken by itself is neither an *A* nor an *E*. For example 'Poisonous gases are colourless' may be treated as an *A*-statement. If one treats it so, then 'Poisonous gases are coloured' is the corresponding *E*-statement. But we could just as well have done it the other way. The point is these two statements are contraries because they affirm contrary terms. The terms *colourless* and *coloured* are logical contraries; the choice of which term is "negative" is relative and not a matter of logic. Quantity has been eliminated. We get the effects usually got by quantity by distinguishing these four ways of predicating a term of a plural subject. Nor does "quality" remain a factor distinguishing *A* and *I* as positive statements from *E* and *O* as negative statements. For *E* affirms the contrary of what *O* denies; if *E* and *O* are "negative", they are so in logically different ways. Because the usual differences of quantity and quality are in this way absorbed by the four modes of predicating a term, I call the above schedule of categoricals "the predicative scheme".

6. The predicative scheme affords a gratifying simplification of immediate and mediate inference. Two rules suffice.

(1) If one term is truly predicated (i.e., affirmed or denied)

of another, the contrary of the second term is truly predicated of the contrary of the first:

$$Ps = \bar{S}\bar{p}; \qquad P's = \bar{S}'\bar{p}.$$

I call (1) the rule of inversion. Assuming that no term has more than one logical contrary this rule suffices for immediate inference. For example, suppose we wished to know whether the converse of E may be immediately inferred from E. We note that $\bar{P}s = \bar{S}\bar{\bar{p}}$ and that $\bar{S}\bar{\bar{p}} = \bar{S}p$ which shows that E is convertible.

Similarly since the obverse of $P's$ is $\bar{\bar{P}}'s$, obversion of O is valid. On the other hand the converse of $P's$ is $S'p$ which is not its inverse. Hence conversion of an O-proposition is not a valid immediate inference.

(2) If one term is truly affirmed of a second and the second is truly affirmed of the third, then the first is truly affirmed of the third:

$$Xy \cdot Yz / Xz.$$

This principle that *affirmative predication is transitive* is the basic principle of syllogistic reasoning. All and only all valid syllogisms are equivalent to arguments having the formal properties of the formula

$$Xy \cdot Yz \cdot / Xz.$$

The formula has three formal properties:

1) all of its statements are affirmative.
2) it is in "transitive" form.
3) it contains no more than three (recurrent) terms.

These properties are necessary and sufficient for validity. We apply them to test the validity of any syllogism. The procedure consists of three steps.

(i) If the syllogism contains denials, transpose them until all statements are affirmative. This can be done only if the conclusion and *one* premise is a denial.
(ii) Next get the implication into transitive form. This can always be done by using inversion.
(iii) Count the terms. There should be no more than three.

Suppose, for example, we wished to test *EIO*.3 for validity. We have

$$\begin{aligned} &\bar{P}m \cdot \bar{S}'m \supset P's \\ \equiv\ &\bar{P}m \cdot Ps \supset \bar{S}m \quad \text{(step 1)} \\ \equiv\ &\bar{M}p \cdot Ps \supset \bar{M}s \quad \text{(step 2)} \end{aligned}$$

The syllogism is valid since there are only three terms, $\bar{M}$, P, and S.

Another illustration. Is *AEE*.1 valid? We have

$$Pm \cdot \bar{M}s \supset \bar{P}s \quad .$$

which is already in affirmative transitive form. It is invalid, however, since it has five, not three terms.

7. The testing procedure is fast and simple. Moreover, it is not an algorithm but a direct logical method. However, it suggests an interesting algebraic algorithm which exploits the following analogies to the rules of inversion and syllogistic:

(A) $Ps = \bar{S}\bar{p}$ corresponds to $S/P = P^{-1}/S^{-1}$.
(B) $Pm \cdot Ms \supset Ps$ corresponds to $M/P \cdot S/M = S/P$.

We now rewrite the categoricals in algebraic form

A. All *A*'s are $B =$ df. a/b.
E. No *A*'s are $B =$ df. a/b^{-1}.
I. Some *A*'s are $B =$ df. $(a/b^{-1})^{-1}$.
O. Some *A*'s are not $B =$ df. $(a/b)^{-1}$.

We need consider only syllogisms all of whose propositions are affirmative or else those with a negative (i.e., particular) conclusion and one negative premise. We may now state (B) as a rule of validity for all such syllogisms.

(B) A syllogism is valid *if and only if* the algebraic product of its premises is equal to the conclusion.

Illustrations: Are *AII*.1; *AEE*.1; *AOO*.2; *EIO*.4 valid?

AII.1
$$\frac{M}{P} \cdot \left(\frac{S}{M^{-1}}\right)^{-1} = \left(\frac{S}{P^{-1}}\right)^{-1}$$

$$\frac{M}{P} \cdot \frac{S}{P^{-1}} = \frac{S}{M^{-1}} \qquad \textit{valid.}$$

AEE.1 $\qquad \dfrac{M}{P} \cdot \dfrac{S}{P^{-1}} \neq \dfrac{S}{P^{-1}}$ *invalid.*

AOO.2 $\qquad \dfrac{P}{M} \cdot \left(\dfrac{S}{M}\right)^{-1} = \left(\dfrac{S}{P}\right)^{-1}$

$$\frac{P}{M} \cdot \frac{S}{P} = \frac{S}{M} \qquad \textit{valid.}$$

EIO.4 $\qquad \dfrac{P}{M^{-1}} \left(\dfrac{M}{S^{-1}}\right)^{-1} = \left(\dfrac{S}{P}\right)^{-1}$

$$\frac{P}{M^{-1}} \cdot \frac{S}{P} = \frac{M}{S^{-1}} \qquad \textit{valid.}$$

8. The logical and algorithmic methods I have outlined are quite generally applicable to any argument using syllogistic reasoning. Nor is there any need to put the argument into "standard form" before applying them. It is, for example, possible to treat the following argument form directly:

Some M is P
All non-S is non-M

Some P is S

We have: $\bar{P}'m \cdot \bar{M}\bar{s} \supset \bar{S}'p$; $\bar{M}\bar{s} \cdot \bar{S}p \supset \bar{M}p$ which shows it is valid. The method may be generalized to include sorites. Any valid sorites of n statements (i.e. $n-1$ premises and the conclusion) must be reducible to a transitive affirmative implication containing exactly n terms. Thus a valid four statement sorites will be equivalent to an implication having the form

$$Wx \cdot Xy \cdot Yz \supset Wz.$$

In the fraction model the ratio of the product of the premises to the conclusion will still be unity. In general if $P_1, P_2 \ldots P_{n-1}$ are the premises and q is the conclusion of an n-statement sorites, then

$$P_1 \times P_2 \ldots \times P_{n-1} = q \equiv P_1 \cdot P_2 \ldots \cdot P_{n-1} \supset q.$$

Illustration: Some clergymen are priests
All priests are bachelors
No bachelors are married

Some clergymen are not married

Direct method

$$\bar{P}'_C \cdot B_P \cdot \bar{M}_B \supset M'_C$$
$$M_C \cdot B_P \cdot \bar{M}_B \supset \bar{P}_C$$
$$M_C \cdot \bar{M}_B \cdot B_P \supset \bar{P}_C$$
$$\bar{C}_{\bar{M}} \cdot \bar{M}_B \cdot B_P \supset \bar{C}_P$$

Valid

Fraction method

$$\left(\frac{C}{P^{-1}}\right)^{-1} \cdot \left(\frac{P}{B}\right) \cdot \left(\frac{B}{M^{-1}}\right) = \left(\frac{C}{M}\right)^{-1}$$
$$\left(\frac{C}{M}\right) \cdot \left(\frac{P}{B}\right) \cdot \left(\frac{B}{M^{-1}}\right) = \left(\frac{C}{P^{-1}}\right).$$

Valid

Enthymemes may be handled by "solving" for the missing premise. This may be done by either method. If, say, the third premise in the above argument is missing we may use the fraction model to find it.

$$\left(\frac{C}{P^{-1}}\right)^{-1} \cdot \left(\frac{P}{B}\right) \cdot \left(\frac{X}{Y}\right) = \left(\frac{C}{M}\right)^{-1}$$
$$\left(\frac{C}{M}\right) \cdot \left(\frac{P^{-1}}{C}\right) \cdot \left(\frac{P}{B}\right) = \left(\frac{Y}{X}\right)$$
$$\left(\frac{Y}{X}\right) = \left(\frac{M^{-1}}{B}\right)$$
$$\left(\frac{X}{Y}\right) = \left(\frac{B}{M^{-1}}\right); \quad \left(\frac{X}{Y}\right) = \left(\frac{M}{B^{-1}}\right)$$

9. *Existential predications.* In the passage I quoted, Russell alludes to the elimination of existence as a predicate as one of the triumphs of the quantificational interpretation of categorical propositions. And indeed it is not easy to see how a statement like 'Tigers exist' can be construed as an "unquantified" predicative statement without falling into musty traps. We know on the other hand that Aristotle was suspicious of existence as a predicate; he, nevertheless, considered existential statements to be simple predications. This raises the question how to construe existential statements "predica-

tively" without the existential quantifier and without treating "exist" as a predicate term.

Let $|P|$ stand for the term "things that are either P– or –$\overline{P}$". The following equivalences then hold.

$P_{\|P\|}$	$\text{E}xPx \cdot \sim \text{E}x\overline{P}x$	(All) things are P
$\overline{P}_{\|P\|}$	$\text{E}x\overline{P}x \cdot \sim \text{E}xPx$	(All) things are $\overline{P}$
$\overline{P}'_{\|P\|}$	$\text{E}xPx$	(All) things aren't $\overline{P}$
$P'_{\|P\|}$	$\text{E}x\overline{P}x$	(All) things aren't P

The predicative scheme for existential statements has existence (or "things") as a logical subject. The term $|P|$ is assumed to be non-empty; if nothing either is or fails to be P, then P is an impredicable term. If, for example $P =$ healthy prime (number) then it will be true that $|P|$ is empty and that $\sim \text{E}xPx \cdot \sim \text{E}x\overline{P}x$. For nothing is a healthy prime or an unhealthy prime or a healthy non-prime or an unhealthy non-prime. The term 'healthy prime' is then said to be impredicable in every domain. The use of logical contrariety as distinct from statement negation allows us to express such ontological truths as "numbers are neither coloured nor colourless". This gives us a tool for type analysis and it is useful for cases of *special* impredicability, for example, in showing that "false" is impredicable in "this statement is false". For *true* and *false* are logical contraries and we get, by the usual reasoning,

$$F's' \equiv \overline{F}'s'$$

$$\text{since} \quad \sim (F's' \cdot \overline{F}'s')$$

$$\text{we get} \quad \sim F's' \cdot \sim \overline{F}'s'$$

which shows that 's' is neither true nor false or. in other words, that *false* is impredicable of 's'.

Note that if $\overline{F}'s'$ is taken as equivalent to $\sim F's'$ we lose the solution. Here again the difference between statement negation and predicate contraffirmation plays a crucial logical role in a correct understanding of the paradox.

10. We do generally assume that both terms of a predicative tie are predicable terms. Thus if P is a term we assume that $\overline{P}'_{|P|} \vee P'_{|P|}$

or in more familiar quantificational formulation – that $\mathrm{E}x(Px) \vee \vee\, \mathrm{E}x(\bar{P}x)$. Moreover, if P and S are in a predicative tie we make the further assumption that both are predicable of the very same things, that both have the same "universe of discourse" in the statement. This is assumed in the rule of inversion or in classical contraposition. It may not actually be true if each term is taken separately. For example it may be the case that whatever wears trousers interests Mary. We may represent this as Iw. It is however not true that $Iw \equiv \overline{W}_{\bar{I}}$ – what fails to interest Mary fails to wear trousers – since twin prime numbers fail to interest Mary but they neither wear trousers nor do they fail to wear them. However, "the universe of discourse" of any statement Ps is confined to things that are $|P|$ and $|S|$. And on *this* assumption $Ps \equiv \bar{S}\bar{p}$. What fails to interest Mary inside the range of things that wear or fail to wear trousers do fail to wear trousers so that Iw is equivalent to $\overline{W}_{\bar{I}}$.

A syllogistic argument with P, M and S is confined to the universe of things that are $|P|$ *and* $|M|$ *and* $|S|$. It is possible to formulate valid syllogisms with significant true premises and an impredicable conclusion. For example:

Nothing that wears trousers interests Mary
Some theorems interest Mary

Some theorems fail to wear trousers

This syllogism is formally valid (since we have $\bar{I}w \cdot (\bar{I}'t) \supset W't$) but the premises clearly outrun the assumption that all terms have the same predicability range for the argument. The terms T and W have nothing in common of which both are predicable and $|T| \cdot |W| \cdot |I|$ is here empty.

11. Any syllogism, then, with terms M, P, and S has the range common to all, i.e. the things that are $|M|$ and $|P|$ and $|S|$. Let these things be E. I wish now to show how arguments like

Humans are mortal
Humans exist

Mortals exist

may be represented as syllogistic in form. To do this we use E as the logical subject:

$$M_H \cdot \bar{H}'_E \supset \cdot \bar{M}'_E \text{ is a valid syllogism.}$$

A statement like $\bar{M}'_E$ may be read as "things aren't (all) M" which is equivalent to saying (non-predicatively) "there are M's". It is therefore possible to include existential statements in syllogistic reasoning. I have not found Aristotle does this. It is however wrong to say that he never gives examples of syllogisms that include *singular* statements. There occur in Aristotle syllogisms like

No mule is pregnant
that is a mule

therefore that is not pregnant.

And of course the medieval Aristotelian considered

All men are mortal
Socrates is a man

Socrates is mortal

to be an example of a perfect syllogism.

The predicative scheme I have outlined does not classify syllogisms according to mood. In that scheme all predications, singular or plural, have unquantified subjects. But this means that singular and general statements have the same logical form. Singular statements therefore enter into syllogisms on a par with general statements; the above argument is a good instance of a transitive affirmation. There is no reason whatever to deny its syllogistic status.

In saying that singular and general predications have the same logical form I do not mean to imply there are no logical differences between them. One fundamental difference was noted by Aristotle: when the subject term, S, is singular then affirming P of S is logically equivalent to denying its contrary and denying P is logically equivalent to affirming its contrary. That is, for singular S,

$$Ps \equiv \bar{P}'s; \qquad P's \equiv \bar{P}s.$$

For example, Socrates is poor $\equiv$ Socrates isn't rich, and Socrates isn't poor $\equiv$ Socrates is rich (assuming that *rich* and *poor* are logical contraries so that *poor* = *un-rich*).

The distinctions between affirmation and contradenial and between denial and contraffirmation come into their own only when the subject is plural and *then* these distinctions are crucial.

We may even use the collapse of these distinctions to show that a statement is singular. The statement "Greeks are numerous" looks like a plural statement. But we note that "Greeks are numerous" is logically equivalent to "Greeks aren't scarce"; this shows the statement is singular. An argument like

Greeks are numerous
Spartans are Greeks
―――――――――
Spartans are numerous

is thereby seen to have four terms since *Greeks* in the minor premise is a general term whereas it is singular in the major [1].

12. In all these applications of the predicative scheme I have been at some pains to distinguish between predicative denial and statement negation. I have not made much actual use of this fundamental distinction; in most of the applications it would be harmless if one were to substitute '$\sim Ps$' for '$P's$'. Nevertheless – and at the risk of having annoyed the reader – I have avoided the more familiar form for a decisive reason. The predicative scheme treats all categoricals as statements of subject-predicate form. Logically compound statements are not predications. Since negations are logical compounds, they too are not predications.

That negations are non-predicative may be seen by the following argument. Neither the statement "The Equator is clean" nor the statement "The Equator is unclean" is a true statement. It follows that their negations are true so that $\sim C_E$ and $\sim \bar{C}_E$.

Suppose now that the subject of '$\sim C_E$' is E. Then '$\sim C_E$'

[1] The observant reader may wonder why the syllogism '$\text{Mortal}_{\text{men}}$ $\text{Man}_{\text{Socrates}} \supset \text{Mortal}_{\text{Socrates}}$' avoids the same fallacy. But a term in predicate position is always general.

denies C of E. Also if E is the subject, '$\sim C_E$' is a singular predication. But in the case of singular predications, denial is logically equivalent to affirming the contrary. It follows that $\sim C_E \equiv \bar{C}_E$ which contradicts the assumption that $\bar{C}_E$ is not a true statement. It is clear that a negation is like any other compound statement in having neither a subject nor a predicate. Of course if '$\sim Ps$' is treated as if it were '$P's$' then s is the subject term. But the negation sign is not a logical sign for denying a term. What *term* is denied in '$\sim$ (Plato is wise or Socrates is bald)'? Similarly one may well ask for the subject and predicate of 'It is not the case that Socrates is bald'.

We were earlier concerned to exploit the logical difference between denying a term and affirming its contrary. What is now at issue concerns another distinction, that between predicative denial and contraffirmation on the one hand and statement negation on the other. In Frege's system both contraffirmation and denial are reduced to some form of statement negation. If all gainsaying is propositional, predication has no place in a logical system. And indeed, in Frege's system, general statements are non-predicative. Where 'Mortal (men)' affirms *mortal* of men, $(x)(\sim \text{mortal } x \vee \text{man } x)$ does not affirm it at all. Quine's extension of quantification to singular statements completes the process of eliminating the subject-predicate distinction from logic altogether. But even those logicians who balk at taking this extreme but logical step fail to realize the implications of accepting negation as a way of representing predicate denial. Removing denial as a distinct predicative mode leaves only affirmation. As affirmation is not contrasted with anything, predication itself is trivialized and rendered mysterious. Frege rightly saw that he could now dispense with the subject predicate distinction as being of no logical importance.

Affirmation and denial of terms gives way before the assertion and negation of whole statements. Of course this gives rise to odd problems, odd doctrines of logical form and odd theories about saturated and unsaturated expressions. But the harm is done. Predication – without denial – is mysterious and logically pointless. Three related doctrines have conspired to reduce predication theory to its present state:

(1) The doctrine that predication is logically singular.
(2) The quantificationalist interpretation of general categoricals as exhibiting their distinct logical form.
(3) The absorption of predicate denial by statement negation.

The first doctrine rules out a predicative interpretation of the four general categoricals and leads into the second by forcing logicians to distinguish between the logical forms of singular and general statements. The second doctrine offers logicians the opportunity to get away with this by enabling them to interpret general statements as essentially quantified and essentially non-predicative.

The third doctrine is an outright dismissal of predication. It dismisses affirmation and denial of terms and replaces it by assertion and negation of statements. In this paper I have dealt mainly with the doctrines and the effects of (1) and (2) but the effects of (3) are perhaps more serious for the foundations of logic as we know it today.

In concluding I wish to turn back to the first doctrine. In combating the dogma of singular predication I offered the suggestion that in the statement 'Men are Mortal' *mortal* is predicated of men and not of each man. The question may seem to arise What sort of entity is "men"? Certainly it is not the class of men that is mortal. Nor am I saying of each man that he is mortal. What is this thing called "men"? This question was asked me by Nelson Goodman who further complained that this use of "men" as the subject of predication went even beyond Plato in its commitment to a new sort of entity. For Goodman and other nominalizers, this is far more than too far. To countenance the class of men is bad enough, to countenance a new thing called men is worse.

This question only seems to arise. I neither "countenance" nor fail to countenance any such thing as men. Only someone who cannot free himself from the dogma of singular predication would even try to formulate the question "What sort of an entity is "men"?" Here the ungrammaticalness of the question attests to its incoherence. Men do not constitute an entity or even a class of entities. I cannot answer the question 'What sort of entity is men?'. I *can* give an answer to 'What are men?'. In the context of ontological commitment the answer may just as well be 'Men are men'. To say – as I have – that *mortal* is true of *men* does nothing to anyone's ontology. The fact that more than one man exists is of zoological, not of ontological interest.

If this seems too flippant an answer let me add the following reminder. Those who follow Quine in using the quantification criterion for ontological commitment must surely find themselves at a loss to say how I can possibly have made anything like an ontological commitment. I have not quantified over any non-individual because I have not quantified at all. I have instead tried to develop the point of view of a philosopher who refuses to quantify subject predicate statements. If I am right this point of view is an interesting perspective on some very old topics.

DISCUSSION

L. Kalmár: *Not Fregean and not a dogma.*

Professor Sommers' 'Fregean dogma' is neither Fregean, nor a dogma. It is not Fregean because it goes back at least to Kant. It is not a dogma because the expression of 'Men are mortal' as '(x) (If x is a man then x is mortal)' is *not obligatory.* There are many ways to express this proposition, which are suited to different purposes. Thus we can express it as '(x) (x is mortal)' provided our universe, the range of values of our variables, is taken to be all men. Or we can express it in the 'Fregean' way, which has the great advantage that we need only one universe, or range of values for our variables. Or we might express it as '(x) (If x is a man then there is a y such that y is mortal and x is identical with y)'; this formulation may be the best for certain purposes.

Sommers has given us another way to express 'Men are mortal'. But his method has many disadvantages: it is not general enough, for it cannot deal with propositions about several subjects, that is, with propositions asserting that a relation holds between several things.

Sommers' algorithm for syllogisms is interesting – but there are other algorithms which also do not depend upon 'Frege's dogma' (for example, Venn diagrams, and the method explained in Hilbert–Ackermann, *Principles of Mathematical Logic*).

M. Dummett: *A comment on 'On a Fregean dogma'.*

Paragraphs 1–8 of *On a Fregean Dogma* set out Professor Sommers' basic thesis, while paragraphs 9–12 introduce certain supplementary doctrines barely touched on in the earlier paragraphs: for convenience, I will therefore refer to paragraphs 1–8 as Part 1, and paragraphs 9–12 as Part 2.

1. Consider the conversion

$$\frac{\text{No horses are blue}}{\text{No blue things are horses.}}$$

This, according to Sommers' analysis, is of the form

$$\frac{\text{Horses are (all) non-blue}}{\text{Blue things are (all) non-horses.}}$$

I claim, first, that there is nothing in Part 1 of the paper which precludes the interpretation of the term 'not-horses' as meaning 'cows' (more strictly, the interpretation of the operator 'not-P' as yielding the term 'cows' when applied to the term 'horses'); and, moreover, I claim that what Sommers says in Part 1 actually debars him from ruling out this interpretation.

2. Let us set on one side for the moment the question whether my claim is sound or not, and ask what, if it is sound, is the objection to Sommers' theory. Going on Part 1 alone, one would think that, assuming the soundness of my claim, the objection would be that an interpretation was permitted under which a supposedly valid inference would lead from a true premiss

Horses are (all) not-blue

to a false conclusion

Blue things are (all) cows.

But Part 2 introduces new considerations, connected with the universe of discourse, and the distinction, not used in Part 1, between 'P'_S' and '$\sim P_S$'. The doctrine advanced in Part 2 would save the above inference from invalidity, even on the interpretation of 'not-horses' as 'cows'. For Sommers maintains the doctrine that, in any given inference, the universe of discourse (range of generality) is to be taken to be the intersection of all classes $|P|$ for every term P occurring in the inference. Thus, in the above inference, the universe of discourse would consist of all those things which are either blue or not-blue, and are also either horses or not-horses. If 'not-horses' is interpreted to mean 'cows', then the conclusion

Blue things are (all) cows

has to be read as meaning that all of those blue things which are either horses or cows are cows: and this is true, and follows from the premiss.

The upshot so far would seem to be that, *if* my claim that 'not-horses' may be interpreted as 'cows' is correct, then, taking into account the doctrine of Part 2, the correct objection to Sommers's theory is *not* that it validates what are in fact invalid inferences, but that it makes the interpretation of a statement depend on what it is inferred from in an obviously intolerable way. This objection will, however, later have to be revised again.

3. I now set out the grounds of the claim I made in (1). Part 1 of the paper advances the thesis that, at least in considering a very restricted fragment of logic, it is possible to dispense with the distinction between singular and general terms, i.e. to avoid taking 'individual predication' as basic, and what we might call 'general predication' as to be explained in terms of it. Now in any system which distinguishes individual from general predication, the operator 'not-P' can be explained as forming from a predicate 'P' a predicate which is, with respect to individual predication, its contradictory. That is to say, although (using Sommers's notation) a statement '$\overline{P}_S$' will in general be only a contrary, not the contradictory, of 'P_S', a statement '$\overline{P}(a)$' will be the contradictory of '$P(a)$' (where 'a' is a singular term). That is, we shall have

$$\text{For every } x, (\overline{P}(x) \text{ if and only if } \sim P(x)).$$

If one wishes not to distinguish individual from general predication, this characterisation of the operator 'not-' becomes unavailable: accordingly Sommers fell back, in Part 1, on stipulating that 'not-P' should be the contrary of 'P'. If the word 'contrary' is taken in its usual sense, 'cows' is a contrary of 'horses'. In the discussion of his paper, however, Sommers explained that he had not meant 'contrary' in this usual sense, but in a special sense which he expressed by saying '*logically* contrary'. The question arises, however, whether, within the limitations which Sommers has imposed on himself, he would be able to explain what this sense is.

4. Going on Part 1, one would think not: for, as I have indicated, the intended interpretation of 'not-' can hardly be explained without invoking the notion of individual predication (or, equivalently, that of a singular term). Part 2, however, suggests that Sommers would after all be able to employ some version of the usual explanation. For in Part 2 Sommers recognises that it is after all necessary, in order to give an account of those inferences which he wants his theory to cover, to distinguish between singular and general terms. At this point one is, indeed, somewhat bewildered as to Sommers' intentions, since the whole thesis of Part 1 had appeared to be that one need not treat singular statements differently from general ones; but I am not enquiring so much into what Sommers' theory achieves, as into whether, as a theory, it is coherent.

Once Professor Sommers has admitted the necessity for distinguishing between singular and general terms, then he would be able to explain 'not-P' in the usual way, viz. as the contradictory of 'P' relative to individual predication. However, another doctrine introduced in Part 2 rules this out, since Sommers advances the thesis that, when 's' is a singular term, 'P'_s' is equivalent to '$\overline{P}_s$', but that neither is equivalent to '$\sim P_s$'. By this means the standard explanation of 'not-P' is rejected as incorrect, i.e. as not giving the interpretation of 'not-P' which Sommers has in mind. The question is thus still wide open *what* explanation of 'not-P' (the operation of forming the 'logical' contrary of a term) Sommers can give, which would rule out the interpretation of 'not-horses' as meaning 'cows'.

5. I illustrate these points from Sommers' remarks about 'true' and 'false'. Sommers wishes to hold *both* that there are statements which are neither true nor false, *and* that 'false' is the ('logical') contrary of 'true', i.e. that 'false' = 'not-true'. Now, on the standard interpretation of 'not-', 'not-true' = 'false' only if every statement is either true or false. It is this which convinces me that my original claim, that Sommers cannot rule out the interpretation of 'not-horses' as 'cows', is correct. For consider this parallel case. We

should normally allow the conversion

No true statement is boring
―――――――――――――――――――
No boring statement is true.

Sommers analyses this as

True statements are (all) interesting
―――――――――――――――――――――――
Boring statements are (all) not-true

(writing 'not-boring' as 'interesting'). Someone who learned that Sommers took 'not-true' to mean 'false', and that he also held that there are statements which are neither true nor false, would, as a first reaction, object that, on this interpretation and under this assumption, the inference was invalid, since there might be boring statements which were neither true nor false. Sommers' only escape, given all that he has committed himself to, would be to plead his 'universe of discourse' doctrine: viz. that, in this inference, the universe of discourse is restricted to those statements which are either true or not-true, i.e. either true or false.

Since the situation with the horses-cows inference seems exactly the same as that with the true-false inference, I conclude that my opinion about the interpretation of the former is correct. Sommers' reply to this must be that the relation of logical contrariety between terms subsists independently of context, and is simply such that, while 'true' and 'false' are contraries, 'horses' and 'cows' are not. If we accept this retort, then my claim in (2) that the interpretation of a statement will depend on what it is inferred from will collapse, when applied to immediate inferences. If we know that 'true' is the contrary of 'false', then we shall interpret the sentence 'Boring statements are (all) false', quite independently of context, as saying that all those boring statements which are either true or false are false, and as therefore not falsified by the existence of boring statements which are neither true nor false. My point about context-dependence will still hold, however, for inferences with more than one premiss. For instance, it may be that there are some statements which, though boring, are admirable, and this would normally be taken to falsify the statement

Boring statements are (all) contemptible.

If, however, none of these admirable but boring statements were either true or false, the truth of the pair of statements

False statements are (all) contemptible
Boring statements are (all) false

would not, given Sommers' 'universe of discourse' doctrine, be impugned; yet the former statement would follow syllogistically from the latter pair.

6. In any case, the problem what Sommers means by 'logical contrary' remains unresolved; and at this point I am frankly uncertain just what idea it is that Professor Sommers wishes to convey by this expression, or whether, indeed, there is any clear idea there to be conveyed. That is, I do not know what that notion of a contrary predicate can be under which it is true to say:

(i) Every predicate has one and only one contrary;
(ii) 'False' is the contrary of 'true';
(iii) There are statements which are neither true nor false;
(iv) 'Cow' is not the contrary of 'horse'.

If there is such a notion, I do not see that Professor Sommers has given the faintest indication of how it is to be explained, or even what it is.

C. LEJEWSKI: *The logical form of singular and general statements.*

Are the two propositions (i) 'Socrates is mortal' and (ii) 'All men are mortal' of the same logical form or not? Professor Sommers argues that they are. The generally accepted view, deriving from Peano and Frege, and supported by Russell, is that they are not. But surely this is not the sort of question that can be answered by simply saying 'yes' or 'no'. Our answers will vary depending, first, on the logical language which we may wish to choose as our 'yardstick' or 'system of coordinates' and, secondly, on the 'depth' of our formal analysis of the propositions under consideration. Thus, for instance, if we construe (i) as being of the form '*Fa*', where '*a*' stands for a singular referential name and '*F*' for a proposition-forming functor for one nominal argument then, with respect to the Frege–Russellian logical language, (ii) cannot be construed as exhibiting the same logical form, simply because it contains no occurrence of a singular referential name. However, with respect to a logical language which, like the one constructed by Leśniewski, makes no formal distinction between singular names and common nouns, both (i) and (ii) can be said to be of the form '*Fa*' with '*a*' exemplified by the nouns 'Socrates' and 'men', and '*F*' exemplified by the proposition-forming functors '. . . is mortal' and 'All . . . are mortal'. Alternatively, and with respect to the same type of language, one could construe (i) and (ii) as exhibiting the form '*Gab*' with '*a*' and '*b*' standing for nouns or noun-like expressions such as 'Socrates', 'men' and 'mortal', and '*G*' – for proposition-forming functors such as '. . . is (a, an) . . .' and 'All . . . are . . .'.

The latter way of construing (ii) can, following Łukasiewicz, be applied to the remaining categoricals, but Sommers seems to favour the former. For him the four categoricals are all of the form '*Fa*'. Accordingly, a syllogism in *Barbara* could be construed as being of the following form: (iii) $Xu \cdot Yz \cdot \supset Xz$. As a formal analysis of *Barbara* this formula is not 'deep' enough to exhibit the universal applicability of the syllogism, and it is easy to think of propositions which exemplify the formula but are false. In fact, Sommers' formula is (iv) $Xy \cdot Yz \cdot \supset Xz$, which appears to indicate that 'Y' is to be understood as a kind of function of 'y'. Informal comments suggest that the expression symbolized by 'Yz' is intended to be equivalent to 'all z's are y's'. But even with this proviso (iv) is not deep enough to resist falsification. This means that a universally true proposition of the form '$Xy \cdot Yz \cdot \supset Xz$' is universally true owing to certain features which are not exhibited in the formula. Once this is realized, any hope of achieving a 'gratifying simplification' by construing the categoricals as instances of '*Fa*' begins to fade away, and if for some reason or other we want to avoid quantification then we can hardly do better than follow Łukasiewicz and construe the categoricals as instances of '*Gab*'.

Sommers' 'predicative scheme' does not appear to have many formal virtues but its contents are of some interest, so it seems. The scheme is based on two primitive notions, that of a term's being truly affirmed of another term and that of the contrary of a term. If we agree that a term is truly affirmed of another term if it stands in the place of 'b' in a proposition of the form 'all a's are b's', and if we symbolize the contrary of a term 'a' by '$\bar{a}$' then by putting 'Aab' for 'all a's are b's' we can express the presuppositions of the scheme as follows:

$$\begin{array}{ll} \text{A1.} & Abc \cdot Aab \cdot \supset Aac \\ \text{A2.} & Aab \supset A\bar{b}\bar{a} \\ \text{A3.} & Aa\bar{\bar{a}} \\ \text{A4.} & A\bar{\bar{a}}a \end{array}$$

The remaining categoricals can be defined by means of the following equivalences:

$$\begin{array}{ll} \text{D1.} & Eab \equiv Aa\bar{b} \\ \text{D2.} & Iab \equiv \sim(Aa\bar{b}) \\ \text{D3.} & Oab \equiv \sim(Aab) \end{array}$$

A1, which is *Barbara*, can be regarded as showing that the notion of a term's being truly affirmed of another term is transitive. It corresponds to Sommers' Rule II. A2, A3, and A4, between them, seem to make explicit the contents of Sommers' Rule I. The system determined in this fashion is a part of the traditional logic. It leaves out all those syllogistic laws in which the premises are all universal while the conclusion is particular. All such laws fail to satisfy the conditions of the algorithm suggested by Sommers.

W. V. QUINE: *Three remarks.*

Professor Sommers writes: "We readily grant that 'are mortal' is true of men only if 'is mortal' is true of each man. But . . . we maintain a sharp distinction between what 'Men are mortal' "really says" and the conditions of its truth." If he maintains a sharp distinction, he should divulge it and drop the disclamatory quotes.

As for ontological commitment, I agree that we are free to couch our sentences in idioms of the logic of quantification, which are quite literally ontological, or in idioms which are not. The latter choice does not yield negative answers to ontological questions, but it does dismiss them.

Meanwhile, independently of dogma and ontology, we can all take interest in efficient logical techniques of other than quantificational form, even if, like Venn diagrams, they work only for monadic cases. But Sommers' model in terms of fractions and reciprocals gives pause, since, if we may use the arithmetic, it equates 'All *A* are *B*' with 'Some *B* are not *A*'. But I suppose he has a device for restraining this effect.

F. Sommers: *Reply.*

'On a Fregean Dogma" is part of a conservative effort to restore categorical statements to predicative status. Two traditional (pre-Fregean) notions – contrariety and denial – are given notational recognition in the "predicative scheme". From critics' remarks it is evident that these notions require more direct attention than they received in the body of the paper. I shall discuss them before going on to detailed points of criticism.

A singular statement of the form (1) '*S* is not *P*' is logically ambiguous. Traditional (Aristotelian) logic treated the expression 'is not' as a sign of term denial, indicating that, in this statement, *P* is denied of *S*. On this interpretation, the predicate expression of '*S* is not *P*' is 'is not *P*' and the word 'not' is part of the negative copula – the "ain't" of predication.

In contrast to this, the average contemporary logician treats (1) as the negation of (2) '*S* is *P*'. In this reading, the word 'not' is a propositional connective having no part in a predicate expression. The contemporary (propositional) interpretation of (1) (which goes back to the Stoics) is logically odd in several respects.

(a) Whereas S is the subject term of (2), it is not the subject term of (1).

(b) Whereas (2) is an elementary proposition, (1) is logically compound.

(c) On the propositional interpretation, affirmation is the only way in which any term can be *predicated*. What we call "denial" is not predicative at all. (As a propositional connective, negation is not predicative. Also it is eliminable in favor of other propositional connectives.)

If 'not' is a propositional sign then (1) is not a proposition of subject-predicate form. In section 11 I asked "what is the subject of 'It is not the case that Socrates is wise'?" Frege himself did not concern himself with this question since he believed the subject-predicate distinction to be of no logical importance. But whether or not one cares about the distinction, it is clear that the term *Socrates* cannot be the subject of '$\sim$ Socrates is wise'. If we do construe this statement as being of subject-predicate form, there are two possibilities:

(i) That Socrates is wise is not the case.

(ii) That not-(Socrates is wise) is the case.

Neither (i) nor (ii) has *Socrates* as its subject term. (i) does allow for denial when being the case is what is denied. Certainly Frege more than any single logician is responsible for the current propositional treatment of 'not' in (i). It is therefore not surprising that he preferred (ii). In preferring (ii), Frege succeeds in eliminating predicative denial even for the single predicate he does consider important, namely 'is the case'.

Considerations of economy leave the contemporary logician with propositional negation as the only logically distinct way of nay-saying. The economy is considerable. The predicative 'not' cannot be used to contradict a compound proposition. Moreover, even where predicative denial does apply, as in (i), there *seems* to be no difference anyway between the predicative and the propositional reading. For if we read (i) as claiming (predicatively) that 'isn't P'

is true of S, then $\sim (S \text{ is } P)$ follows anyway. That this is so can be seen from Aristotle's predicative formulation of the law of contradiction: $\sim (Ps \cdot P's)$. This formulation applies only to elementary propositions. It therefore lacks the generality of $\sim (p \cdot \sim p)$ where p can be *any* proposition. Nevertheless, it serves to show that we can infer '$\sim Ps$' from '$P's$'. What then is the difference between propositional negation and predicate denial?

The answer is that the *converse* entailment does *not* hold in Aristotelian logic. In *De Interpretatione*, Chapter 10, Aristotle states the important equivalence between *denial* and *contraffirmation*. To deny *wise* of *Socrates* is equivalent to affirming *unwise* of *Socrates*. This equivalence holds for singular statements (see section 10) and it may serve as a condition for defining contraffirmation in terms of denial so that if P and Q are logical contraries then $(x)\ [(P'x \equiv Qx) \cdot (Q'x \equiv Px)]$. In the *Categories*, Chapter 10, Aristotle says that if Socrates never existed the *negations* of 'Socrates is ill' and 'Socrates is well' are both true. It follows that whereas *denying 'ill'* is equivalent to affirming 'well', negating 'Socrates is ill' is not equivalent to affirming *well* of *Socrates*. It follows also that in Aristotelian logic the schema

$$\sim Ps \cdot \sim P's$$

is not a contradiction since where Socrates does not exist we do have

$$\sim \text{Socrates is well} \cdot \sim \text{Socrates isn't well} \cdot \sim \text{Socrates is ill}$$

(I use the contraction for predicate denial.) Aristotle also permits this in statements like

(1) Robert Kennedy is a future president.
(2) Robert Kennedy isn't a future president.

In considering such pairs of statements neither true nor false, Aristotle is simply applying his own *predicational* definitions of truth and falsity according to which an S–P statement is true if it affirms P of what-is-P or denies P of what-is-not-P. Since Robert Kennedy is neither what-is-a-future-president nor what-is-not-a-future-president [1] (any more than a non-existent Socrates is ill or well) such statements are – predicatively speaking – neither true nor false. Propositionally speaking, the negations of both are true and both (1) and (2) are propositionally false.

A predication is a statement affirming *or denying* one term of another. Clearly quantified statements are not predications in this sense. In his article "Ifs and Cans" Austin observes that 'If anything is a man then it is mortal' is a conditional and not a categorical statement. Austin finds this to be a peculiar way of construing 'All men are mortal'. Whether or not it is "peculiar", the point is well made. The so-called categorical statements of subject-predicate form are neither predicative nor categorical in the quantificationalist interpretation. In sections 1–5 I argued that the "dogma of singular predication" is responsible for the non-predicative treatment of categoricals. In section 2 the source of the dogma is located in the contemporary interpretation of predicative denial as propositional negation. Once we are convinced that 'X isn't P', 'X is un-P', and '$\sim X$ is P' are all equivalent, the way to a predicative interpretation of the categorical is barred. The predicative scheme distinguishes the categoricals by the four modes of predication. The removal of the differences between denial, contraffirmation and negation (in favor of negation) precludes the predicative interpretation of A, E, I, O statements. It leaves 'is P' as the only predicative mode.

The claim that singular and general statements do not differ in logical form amounts to no more than the claim that both are categorical predicative statements. Mr. Dummett is under the impression that this position must lead to denying all logical differences between singular and general statements or, alternatively, between singular and general terms. He says: "the whole thesis of Part 1 had appeared to be that one need not treat singular statements differently from general ones" but "the intended in-

---

[1] In Aristotle, to predicate is to attribute a determination. If Kennedy is what-is-a-future-P then he has that attribute now and this would be no mere potentiality. Nothing can be done about any attribute he now possesses. If he is what-is-a-future-$\bar{P}$, that too is actual. In either case, an election would be pointless.

terpretation of "not-" can hardly be explained without invoking the notion of individual predication (or, equivalently, that of a singular term)".

Now it is my thesis that the difference between "is" and "are" is not logically basic; we could just as well have used some neutral copula as a sign for affirmation. (In the vernacular, "ain't" is indifferently used for predicative denial in both singular and general statements.) It is not my thesis to "dispense with the distinction between singular and general terms". I do not know why Dummett should think that a simple predicative treatment of *A*, *E*, *I*, *O* statements removes the difference between singular and general terms. Certainly Aristotle held that both general and singular categoricals were predicative; it did not prevent him from formulating the logical differences between them. Perhaps Dummett believes that the confused doctrine of distribution is the inevitable price for construing general categoricals as predicating one term of another. If that were true, the quantificational "translation" which is not predicative but which avoids all confusions about distributed and undistributed terms might well be preferable to a predicative reading. It is however unprofitable to speculate on the source of Dummett's belief that a neutral predicative reading of *A*, *E*, *I*, *O* has the effect of removing logical distinctions between singular and general terms and/or statements. The predicative scheme I outlined requires no distinctions of "distribution". The subject of affirmation, contraffirmation, denial, and contradenial remains the same in all four categoricals. This is achieved by construing the predicative *relation* as indifferent to singular and plural. But to say that both in '*S* is *P*' and '*S* are *P*' the term *P* is predicated of *S*, is not to be indifferent to logical distinctions arising out of the difference between singular and general *terms*.

Dummett's further claim that the notion of logical contrariety must be incoherent in any system that does not distinguish between singular and general terms rests on the same confusion. He concludes this part of his criticism with the remark that I have not given the faintest indication of how to explain the notion of a *logical* contrary. In section 10 I gave more than a faint indication. Once one grants the difference between denying a term (*X* isn't *P*) and negating the statement affirming that term ($\sim(X$ is $P)$), the notion of a logical contrary is clear enough. P and Q are *logical* contraries only if (X isn't $P = X$ is Q) and (X isn't $Q = X$ is P). Thus the condition for logical contrariety is given by Aristotle's predicative equivalence for *singular* statements according to which

$$\text{denial} = \text{contraffirmation}$$
$$\text{affirmation} = \text{contradenial}.$$

The reason 'cow' is not the *logical* contrary of 'horse' (or 'square' of 'circular') [1] is that

$$x \text{ isn't a horse} \neq x \text{ is a cow}$$

Dummett notes that I do not avail myself of what he calls the "standard explanation" of un-P according to which 'un-$Px = \sim Px$'. This doctrine has '$\sim P$' as the "contradictory" (and not merely the contrary) of P. Since I do not know what the contradictory of a term is (the expression '$\sim P$' is simply ill-formed; it is not a term) I did not avail myself of this "explanation".

The contraries *true* and *false* seem to Dummett especially objectionable; it seems to him that 'False$_x$' had better be treated as equivalent to '$\sim$ True$_x$'. He is at some pains to show what difficulties and special pleadings are needed by anyone who forgoes this "standard" definition. In this connection I think it should be noted that even Tarski considered *true* and *false* to be contraries and not "contradictories". For where x is given a value, α, that is not a sentence then 'true$_x$' and 'false$_x$' are satisfied by no objects and we have $\sim$ true$_\alpha$ $\sim$ false$_\alpha$. He therefore qualified his adequacy conditions "T'x' $\equiv X$; F'x' $\equiv \sim T$'x'" by requiring that 'x' be the name of a sentence. In section 9 I point out that the liar's sentence leads to '$F_{\cdot L \cdot} \equiv T_{\cdot L \cdot}$' which has $\sim F_{\cdot L \cdot} \cdot \sim T_{\cdot L \cdot}$ as a consequence. Once we reject the unrestricted equivalence '$T_x \equiv F_x$', thus formally allowing for $\sim T_x \cdot \sim F_x$, this result is significant in showing

[1] Logically contrary pairs like (clean, unclean), (colored, colorless) *do* satisfy the predicative equivalence. The equivalence guarantees that a term can have no more than one logical contrary. Thus if $\bar{P}$ is the (logical) contrary of P and $\bar{\bar{P}}$ is the logical) contrary of $\bar{P}$, then $\bar{\bar{P}} = P$. (See section 6.)

that even certain sentences are outside the range of T-or-F things. To accept – as Dummett does – '$\sim$ *true*$_x \supset$ *false*$_x$' is to be committed to the view that '2 is false' is true and so on – a view that may be found in early Frege.

Dummett foresees elaborate difficulties for those who treat *true* and *false* as contraries. But these are all spurious. Using contrariety we are able to form the referential range of S in a statement in which S occurs in subject position. Thus in 'P_s' P is predicated of those S that are P-or-$\bar{P}$ and in '$Pm \cdot Ms \supset Ps$', M and P are predicated of those S that are P-or-$\bar{P}$ and M-or-$\bar{M}$. And in general all the elementary statements in a compound statement with terms $A, B, C \ldots K$ will be about things in $|A| \cdot |B| \cdot \ldots \cdot |K|$. Dummett's criticism of this doctrine is odd. While he is prepared to consider $|S| \cdot |P|$ a "context-independent" referential range for S as it occurs in 'Ps', he balks at $|M| \cdot |P| \cdot |S|$ as the range for S in its occurrences in '$Pm \cdot Ms \supset Ps$'. For this means that taken by itself 'Ps' is about those S-things that are $|P|$ whereas taken in the second "context" it is about those S that are $|P| \cdot |M|$. I cannot see why Dummett finds this objectionable. It is after all a reasonable and wholly unsurprising consequence of the definition of a range in terms of $|A| \cdot |B| \cdot \ldots \cdot |K|$. Moreover Dummett's own laboured example illustrates the occasional usefulness of such formally relativized ranges. For in the isolated statement 'Boring statements are contemptible' the term *Boring statements* does not refer to boring and |false| statements whereas in 'False statements are contemptible' the same term does refer to boring statements that are |false|. The restriction in the second case is wholly benign: the context is logical and the range is given without reference to meanings of terms. Nor is it any more "contextual" than the restriction of *boring statements* to those that are or fail to be contemptible – a restriction that Dummett calls context independent when applied to the isolated statement 'Boring statements are contemptible'.

Professor Kalmár points out that quantificational interpretations of general categoricals antedate Frege and that Frege's own version of quantification is not the only one. Certainly this is so. But it does not change the fact that all quantificationalist readings of A, E, I, O propositions including the non-Fregean alternatives mentioned by Kalmár operate with singular predication. The idea that predication is logically singular is what I labelled a "Fregean Dogma". Not that Frege was the first to be beguiled by it. But the Frege–Russell doctrine that singular and general categoricals differ in logical form because the latter are quantified is a direct consequence of that dogma.

Professors Kalmár and Lejewski complain that the predication interpretation lacks certain "formal virtues". Kalmár says that relational propositions cannot be covered. Lejewski notes that even syllogistic inference is not fully covered since the weakened moods are not valid in the scheme. As it stands the predicative scheme works for syllogistic inference. (The weakened moods can easily be accommodated.) In this restricted area of logical inference a predicative logic does what quantification does and it does it naturally. The scheme is not meant to be a logical instrument of any generality. Kalmár *may* be right when he says that a predicative model cannot be extended to propositions construed as predicating a relational property of several subjects. But the classical view of 'Socrates is taller than Plato' has *taller than Plato* as the predicate term. It is true that this way of viewing such statements results in a severe loss of inference power. Nevertheless the question of what is being said about what is not necessarily a function of inference power and the classical interpretation does have certain advantages of a semantic kind. Also I am not convinced of the impossibility of constructing a logically useful dequantified scheme for n-ary predicates though the technical difficulties seem formidable.

The question of weakened moods is raised by Lejewski. A syllogism like $Pm \cdot Ms \supset \overline{P}'s$ is not valid in the scheme as I presented it. Of course it isn't valid in a q-system either. But it seems that a predicative system ought to cover all classical inferences. To repair the deficiency we need the following additional postulate (M):

(M) If P is truly affirmed of S, its contrary is truly denied.

$$Ps \supset \overline{P}'s; \qquad \overline{P}s \supset P's$$

$$A \supset I \quad ; \qquad E \supset O$$

Those who find (M) objectionable are under the spell of the view that the subject terms of A and E propositions have no "existential import" while those of I and O propositions do. This view in turn is a consequence of the q-reading of A, E, I, O propositions according to which A and I are translated as '$(x)(S_x \supset P_x)$' and '$Ex(S_x \cdot P_x)$'.

I have said nothing about existential import since nothing in the predicative scheme offends against considering *all four* categoricals as having it or as lacking it. On either alternative (M) is unobjectionable and I should wish to add it to the tools for testing validity. The above syllogism of form $AAI.1$ is now easily seen to be valid [1] since $Pm \cdot Ms \supset Ps$ and $Ps \supset \overline{P}'s$.

The predicative scheme allows for a unified interpretation of all categoricals with respect to existential import. I would myself prefer to consider A, E, I, and O propositions to be existentially neutral. Thus a statement like ($C'u$) 'Unicorns aren't carnivorous' can be considered true even though there do not happen to be any unicorns.

The q-interpretation is forced to distinguish universal from particular propositions and forced to deny the validity of the entailment from affirmation to contradenial. To be so forced is a defect of the q-interpretation. If 'All S is P' does not entail 'Some S is P' it ought at least to entail its possibility. But even '$A \supset \Diamond I$' and '$E \supset \Diamond O$' must be denied by the quantificationalist. For on the q-reading, it is true that all square circles are hexagonal but *impossible* that some are. Of course those who have trained themselves to such comfortable truths as 'Every unicorn is red and blue all over' will hardly be uneasy at such consequences ar at the fact that even $\Box A$ does not entail $\Diamond I$.

Professor Quine's first and second remarks need little comment. I shall be glad to remove the disclamatory quotes from "really says" and accept this as an erratum. I am however unhappy that Quine finds me refusing to divulge the secret about the "sharp distinction" between what a statement says and conditions which

---

[1] It should however be noted that the fraction model remains restricted to non-weakened moods.

must hold in order for that statement to be true. That distinction is more lacking in bluntness than in sharpness and I am at a loss to know what Quine wants to know.

Quine in his second remark agrees that a quantifier-free interpretation of categoricals makes no ontological assumptions. This is not surprising and requires no comment.

His last remark takes note of the fact that the propositional interpretation of fractions cannot allow for $(a/b)^{-1} = (b)/(a)$. This reflects the fact that '-1' does double duty as a negative particle attached to a

'term' $\left(\frac{(\ldots)}{(\text{——})^{-1}}\right)$, representing contrariety or to a

'proposition' $\left(\left(\frac{(\ldots)}{(\text{——})}\right)^{-1}\right)$, representing its denial.

We cannot therefore allow an external inverse sign to be eliminated by crossing the parentheses enclosing the fraction in the usual arithmetical way. In its use as a syllogistic algorithm this restraining device is already built in. For we consider only those equations that have no external inverse signs (all propositions being universal) or else those with *one* external inverse on *each* side. In an equation like '$P/M \cdot (S/M)^{-1} = (S/P)^{-1}$', representing the valid syllogism *AOO*.2, the external signs are not eliminated by entering inside the fraction but by one another. Thus the algorithm as I gave it is not subject to the ambiguity noted by Quine; the occurrence of one external inverse sign on each side obviates the operation $(a/b)^{-1} = b/a$. If we did not restrict the candidate equations in this way, the ambiguity would indeed affect the usefulness of the algorithm. For example *OOO*.1 would be valid since, arithmetically, $(P/M)^{-1}(M/S)^{-1} = (S/P)^{-1}$.

Kalmár notes that other algorithms are available. An algorithm is no more than a device. The fraction model is fast and easy to use. Moreover it permits us to solve for a missing premise in a valid enthymeme. But, finally, this particular algorithm is a natural analogy to the predicate laws '$Ps = \bar{S}\bar{p}$' and '$Pm \cdot Ms \supset Ps$' and, as such, it well represents the predicative scheme outlined in the body of the paper.

In more fully explaining the predicative scheme, I hope I answered some of Lejewski's more general criticisms about its "depth". It is clear that in a quantifier-free predicative reading of 'All S is P' the terms S and P are of the same syntactical type. This allows for significant convertibility so that if 'S's are P's' is well-formed, so too is 'P's are S's'. It permits us to consider the two 'Y''s in the formula $Xy \cdot Yz \supset Xz$ as two occurrences of the same term despite the fact that it occurs in different predicative positions. Lejewski over-cautiously misrepresents the case when he construes *Barbara* as of form $Xu \cdot Yz \supset Xz$ – a "form" that allows for the syntactic ambiguity of the middle term at the expense of the principle of transitivity of affirmation. When 'S's are P's' is predicatively construed, without the quantifier expression 'all' or 'every', there can be no temptation to consider 'all S' as the subject-term – and no reason to deny that the same term can occupy the subject position in one statement and be predicated in another.

Lejewski says the predicative scheme "is based on two primitive notions, that of a term being truly affirmed of another term, and that of the contrary of a term". This becomes accurate if we substitute "predicated" for "affirmed". But then one might say that the primitive notions are three: affirmation, denial, and contrariety. The first two are ways of predicating terms. The third names the symmetrical relation that a pair of terms (P and Q) bear to one another when singular statements affirming and denying them are analytically equivalent.

I have tried to show that a quantifier-free reading of A, E, I, O statements embodying these three notions is the interpretation that best leaves these statements alone. And when they are left alone they are what they seem to be: categorical statements predicating one term of another.

RECENT RESULTS IN SET THEORY

ANDRZEJ MOSTOWSKI

Warsaw University

The aim of this paper is to review some recent results reached in the metamathematical investigation of set theory and to discuss their relevance to the problems of foundations of mathematics. More specifically we shall try to show that there are several essentially different notions of set which are equally admissible as the intuitive basis for set theory.

The notion of set seems to have never been understood in a unique way by mathematicians. We find in Becker [2], p. 316, an instructive account of a conversation which took place between Cantor and Dedekind. Whereas Dedekind compared sets to bags which contain unknown things, Cantor took a much more metaphysical position: he said that he imagined a set as an abyss.

The following remarks should make it clear that the divergence of opinion about the nature of sets is very important for the foundations of mathematics.

The founders of set theory hoped that it would provide a basis for the whole of mathematics. They wanted to define in set-theoretical terms all the notions of ordinary mathematics, and to prove by means of set-theoretical laws all the theorems concerning these notions. It is well known that this plan can be executed, and that a set-theoretical reconstruction is possible not only of classical mathematics which deals with numbers, functions, points, and geometrical figures, but also of large parts of modern mathematics, as developed for instance by the Bourbaki school. Only the very recent theory of categories contains notions which do not entirely fit in the set-theoretical frame.

The reduction of mathematics to set theory would provide us with a satisfactory basis for mathematics if set theory were a clear and well-understood branch of science. Unfortunately this is not the case. The notion of a set is much more complicated than was originally thought. Various ways of making this notion more precise were proposed during the discussions of the foundations of set theory. This accounts for a multitude of axioms which were proposed for set theory as well as for the fact that none of these axiomatic systems has been unanimously accepted by all mathematicians.

We should also note that many mathematicians, and especially those who work on classical analysis and geometry, completely ignore abstract set theory, remarking, not without some understandable professional maliciousness, that the internal difficulties of this theory have no bearing on the development of mathematics.

A number of general principles concerning sets were accepted almost universally by mathematicians, and these principles exerted a great influence on the development of abstract parts of modern mathematics and on the way in which mathematics is taught. But these principles do not exhaust the whole of set theory and are much too weak to enable us to solve even moderately deep set-theoretical problems.

All this does not diminish the philosophical importance of set theory. The possibility of interpreting within this theory most (if not all) mathematical notions is a remarkable phenomenon which evidently calls for explanation.

This possibility is due to a very specific characteristic of set theory which is not shared by any other known mathematical theory. Most mathematical theories limit themselves to a study of objects of a well-defined type and certain well-defined relations between these objects; the relations are mentioned in these theories but do not belong to their universes. Such is the situation in the case of arithmetic, analysis, geometry etc. In set theory we admit that all sets belong to the universe of the theory as well as all relations which can be defined in a set. Formally this circumstance is expressed by the power set axiom whose use is essential in the proof that all relations with a given field form a set.

There are other more powerful constructions which lead from a set to another more comprehensive set. E.g. we may start with a set x, form the family x^* of all relations with the field x and iterate this operation infinitely many times taking at the limit points unions of the sets already constructed. In justifying this construction in the axiomatic set theory we use not only the power set axiom but also the sum set axiom and the axiom of replacement[1].

The existential assumptions of set theory are thus very strong. Once we realize this we understand why it was possible to interpret so many theories in set theory. It is also clear that it is certainly not easy to find a justification for such strong existential assumptions.

The existential assumptions discussed thus far were formulated and accepted in set theory from its very beginning. In recent years set theoreticians have formulated and advocated several new assumptions of an existential character. These assumptions are known as 'axioms of infinity'. I shall later review some of them.

There are two general principles which allow us to formulate infinitely many such axioms.

The first of them may be called the *principle of transition from potential to actual infinity*. An early application of the principle occurs in Dedekind's alleged proof that there exist infinite sets; see [2], p. 316. Dedekind started from an arbitrary object S_0 and performed on it certain operations which led to new objects $S_1, S_2, \ldots$ Then Dedekind assumed that there exists a set of all these objects. This essential step is based on an assumption which Dedekind considered as self-evident. In axiomatic set theory this step is justified by the use of a special axiom, the axiom of infinity.

In a more sophisticated form the principle of transition from potential to actual infinity is used in the formulation of the axiom of inaccessible numbers. The Zermelo–Fraenkel axioms state that the set-theoretical universe is closed with respect to certain operations and hence that it is 'potentially closed' in a certain sense of this word. According to the general principle we assume an axiom stating that not only the universe but also a set (i.e. an object of the universe) is closed with respect to these operations.

When formulating this axiom carefully one notices that the closure condition with respect to operations described by the

[1] For general information concerning axioms of set theory see e.g. Fraenkel–Bar-Hillel [7]. By axiomatic set theory we mean the system based on the axioms due to Zermelo and Fraenkel.

axiom of replacement can be expressed in a twofold way. We can either require that a set be closed with respect to the operation of forming an image fy of a set y where f is an *arbitrary* function, or that it be closed under this operation only in case when f is a *definable* function. We have thus two axioms corresponding to the stronger or to the weaker closure conditions.

The axiom in the strong form is equivalent to Tarski's axiom of inaccessible cardinals [23]. Montague and Vaught [15] proved that the weak form of the axiom is not equivalent to the strong one although it too is independent of the axioms of Zermelo–Fraenkel.

In a more general form the transition principle was used by Lévy [12]. His scheme of the axiom of infinity says that if $a, b, \ldots, m$ are sets satisfying a set theoretical formula F, then there is a transitive set s which is closed with respect to all the operations described in the Zermelo–Fraenkel axioms, contains $a, b, \ldots, m$ as elements, and has the property that $a, b, \ldots, m$ satisfy F in the set s. Thus Lévy's scheme says that the property of the universe expressed by the fact that $a, b, \ldots, m$ satisfy F is reflected in the set s.

Lévy's scheme is much stronger than the axiom of inaccessible cardinals. Using this scheme we can prove for instance that there exist inaccessible cardinals m such that there are m inaccessible cardinals n satisfying the inequality $n < m$. Such cardinals m form the so-called first class of Mahlo's cardinals. Also cardinals of many other Mahlo classes can be proved to exist on the basis of Lévy's scheme; cf. [12].

Still stronger axioms of infinity can be obtained by the use of the second principle; we shall call it the *principle of existence of singular sets*. This principle, which is much less sharply defined than the previous one, is concerned with the following situation. Let us assume that in constructing sets by means of the operations described by those set-theoretical axioms which we have accepted so far, we obtain only sets with a property P. If there are no obvious reasons why all sets should have the property P, we adjoin to the axioms an existential statement to the effect that there are sets without the property P. In this form the principle is certainly far too vague to be admissible. It is an historical fact, however, that several axioms of infinity were accepted with no other justification than that they conform to this vague principle.

The first application of the principle was due to Mahlo [14] who postulated the existence of ϱ_0-cardinals. He defined ϱ_0-cardinals as weakly inaccessible cardinals m with the property that there exists at least one continuous and increasing sequence of weakly inaccessible cardinals whose limit is m and that each such sequence contains at least one term which is weakly inaccessible. Nowadays we use a different notion of ϱ_0-cardinals whose definition can be obtained from the one given above by replacing everywhere the words 'weakly inaccessible' by 'strongly inaccessible'. Mahlo considered also a whole hierarchy of ϱ_0-cardinals. The existence of all these cardinals requires special axioms which are formed according to the principle of existence of singular sets [1].

Another application of the second principle is visible in the formulation of the axiom stating the existence of measurable cardinals. A cardinal m is called measurable if the family of all subsets of a set of power m contains a non-principal m-ideal (i.e., an ideal which is n additive for every $n < m$). Tarski and Ulam [22], [25] proved already in 1930 that all cardinals smaller than the first inaccessible cardinal are non-measurable. In 1960 Tarski and his collaborators [24] extended this result for many other cardinals, including e.g., all cardinals smaller than the first ϱ_0-cardinal. Bukovský and Příkrý [4] proved a general meta-mathematical theorem in which they exhibited a large class of set-theoretical formulae all of which have just one free variable and have the property that the following formula is provable in the axiomatic set theory: If there are cardinals (i.e. initial ordinals) satisfying $F(x)$, then the smallest such cardinal is not measurable. It follows in particular that the first inaccessible cardinal, the first cardinal of the first Mahlo class, etc., are all non-measurable. In spite of all these results no proof that all cardinals are non-measurable seems to exist. Hence we experiment with an axiom

---

[1] Bernays [3] derived the existence of ϱ_0-cardinals from axioms which have the form of reflexion principles and thus are applications of the first principle rather than the second.

to the effect that there exist measurable cardinals. This axiom was first formulated by Tarski [24]; first applications are due to Scott [20]. An exhaustive bibliography of papers dealing with this axiom and with related questions is contained in [10].

Still stronger axioms can be obtained by introducing a hierarchy of measurable cardinals similar to the Mahlo hierarchy. Another possibility is to investigate stronger properties of ideals than the one we used in the definition of measurability. Thus e.g., Tarski and Keisler [10] considered the property of m stating that every ideal in the field of all subsets of a set of power m can be extended to a prime m-ideal.

The strongest axiom of infinity known to me was formulated by Bukovský and Příkrý [4] but the details of their work are too complicated to be presented here.

While it is not difficult to show the independence of the axioms of infinity, proofs of their relative consistency are as good as hopeless. A straightforward application of Gödel's second incompleteness theorem shows that no such proof can be formalized within set theory. In view of what has been said above about the reconstruction of mathematics in set theory it is hard to imagine what such a non-formalizable proof could look like. Thus there does not exist any rational justification of the strong axioms of infinity.

We shall now discuss the question whether these axioms are relevant for more conservative portions of set theory which deal with sets of limited powers.

We shall concentrate upon the problem of characterizing true existential or conditionally existential statements concerning sets of integers. We shall consider a very limited class of such statements: we shall assume that all quantifiers which occur in these statements are restricted to integers or to sets of integers. We assume of course that there exist the set of all integers and the family of all its subsets since otherwise our question would be meaningless. Because of this assumption we admit as true all statements of the form $(\mathsf{E}X)(k)\ [k \in X \equiv F(k)]$ where F is any formula with at least one free variable k and not containing the variable X. But apart from this simple case very little can be said about the truth of even very simple sentences such as e.g., $(\mathrm{E}X)\ F(X)$, $(X)\ (\mathrm{E}Y)\ F(X, Y)$, $(X)\ (\mathrm{E}Y)\ (Z)\ F(X, Y, Z)$ where F is an arithmetical formula.

We can construct a sentence $(X)\ (\mathrm{E}Y)\ F(X, Y)$ with an arithmetical F, which is not provable in the axiomatic set theory but derivable in this theory from the assumption that all sets have powers smaller than the first inaccessible cardinal. We obtain this sentence by expressing in the language of second order arithmetic the sentence: there is no well-founded model for the Zermelo–Fraenkel axioms. The sentence is obviously false in intuitive set theory although it is undecidable on the basis of the axiomatic theory of sets; cf. [16].

If we allow three quantifiers in the prefix, then we can construct sentences which are undecidable in a much stronger sense. The axiom of constructibility formulated for the first time by Gödel [9] is a case in point. It has been shown by Addison [1] that the sentence: every set of real numbers is constructible, can be put in the form $(X)\ (\mathrm{E}Y)\ (Z)\ \ F(X, Y, Z)$ where F is an arithmetical formula. This sentence is now known to be independent of the axioms of Zermelo–Fraenkel, even if we add to them various strong axioms of infinity. No convincing reasons seem to exist for accepting or rejecting this axiom in intuitive set theory. The general consensus among the mathematicians is that the axiom is probably false. We may note in passing that the axiom of constructibility is incompatible with the axiom stating the existence of measurable cardinals (cf. Scott [20]).

Problems pertaining to the falsity or truth of statements concerning definable sets of sets of integers were systematically treated in the theory of projective sets created by Lusin [13]. Only very few of the problems formulated in this theory and centering around the so-called separation principles were solved. Some of them were solved with the help of the axiom of constructibility (cf. [17] and [1]) but it is not known whether the use of this axiom is essential in the proofs. Most of the questions remain open and it is highly significant that Lusin maintained that 'we shall never know' the answers.

It is very surprising that the use of strong axioms of infinity may be of importance for questions dealing with projective sets.

Rowbottom [19] and Gaifman [8] announced recently that the existence of measurable cardinals implies the denumerability of the family consisting of all constructible sets of integers. This connection between the fantastically large cardinals and a relatively simple arithmetical property of sets of integers (expressed by a formula $(\mathrm{E}Y)(Z)\ F(X, Y, Z)$ with an arithmetical F) is one of the strangest phenomena discovered in the theory of sets.

It is obvious from the above sketchy review that the problem of the intuitive truth of sentences is difficult and not independent from the problem of truth of strong axioms of infinity even if we restrict ourselves to sentences of second order arithmetic. For this reason many mathematicians abandon intuitive set theory in favour of axiomatic set theory. We shall discuss in what follows some metamathematical results reached in the course of this study.

Most important results deal with the relative consistency and independence of various set theoretical hypotheses. I mentioned above the status of these problems with respect to axioms of infinity. The deepest result is of course the relative consistency of the axiom of choice and of the generalized continuum hypothesis established more than a quarter of century ago by Gödel [9]. His proof contains a much deeper result than a mere proof of consistency. He recognized that the intuitive notion of a set is too vague to allow us to decide whether the axiom of choice and the continuum hypothesis are true or false. He therefore searched for a more precise notion of a set and discovered that a transfinite iteration of the predicative set constructions yields a class closed with respect to the basic operations described by the Zermelo–Fraenkel axioms. Hence we can take the notion of a constructible set as a more precise notion which can replace the vague intuitive notion. In the realm of constructible sets the axiom of choice and the generalized continuum hypothesis are valid. This entails in particular that they are consistent with Zermelo–Fraenkel system. The result can be extended to systems obtained from the Zermelo–Fraenkel system by adjunction of some but not all strong axioms of infinity.

We mentioned before that the axiom of measurable cardinals

is incompatible with the assumption that all sets are constructible. The problem whether the axiom of choice is consistent relatively to the Zermelo–Fraenkel system extended by the axiom stating the existence of measurable cardinals cannot thus be solved in the same way as the problem solved by Gödel, and the same is true also for the generalized continuum hypothesis.

In the case of the axiom of choice a solution can be obtained by a slight modification of an unpublished consistency proof proposed by Scott. Instead of constructible sets Scott considered sets definable in terms of ordinals. In order to define this notion we denote by $R(\alpha)$ the family of sets of rank α. A set X is definable in $R(\alpha)$ if there is a formula F such that X is the unique element of $R(\alpha)$ which satisfies F in $R(\alpha)$. A set X is definable in terms of ordinals if, for some α, it is definable in $R(\alpha)$. Scott showed that the axiom of choice is true in the domain of these sets. A slight modification allows us to obtain a consistency proof of the axiom of choice relatively to Zermelo–Fraenkel axioms with the addition of the axiom of measurable sets. It is sufficient to replace the notion of definability in terms of ordinals by that of definability in terms of ordinals and a fixed ideal.

In case of the generalized continuum hypothesis the problem of its relative consistency remains open when the axiom of measurable cardinals is assumed. Solovay [21] showed the relative consistency of the ordinary continuum hypothesis and Příkrý [18] the relative consistency of the continuum hypothesis for the first measurable cardinal. Both these authors use the forcing method due to Cohen [5]. This very powerful method allowed Cohen and his followers to solve almost all independence problems in axiomatic set-theory.

If we disregard the very easy problems concerning independence of the axioms of infinity, then the simplest but already quite deep problem of independence will be that of the axiom of constructibility. We shall limit ourselves to this problem.

Let M be a transitive family of sets such that all the axioms of Zermelo–Fraenkel as well as the axiom of constructibility are valid in M. All elements of M are then constructible sets. We denote by $\alpha(x)$ the ordinal which indicates the place in which a constructible set occurs in the transfinite sequence of the predicatively defined sets. Since the axiom of constructibility holds in M the ordinal $\alpha(x)$ is in M for every x in M. Let N be another transitive set such that all axioms of Zermelo–Fraenkel are valid in N and such that M is a proper subset of N but M and N have the same ordinals. The axiom of constructibility is then false in N since for an x in $N-M$ the ordinal $\alpha(x)$ is not in N; otherwise $\alpha(x)$ would already belong to M and hence so would x.

The independence will thus be proved when we show that there exists a model N which is a proper extension of M and has the same ordinals.

The construction of models for set theoretical axioms is difficult in view of the complicated non-predicative character of axioms and especially of the power set axiom and the two axiom schemes: the scheme of set construction (which states the existence of the set $\{x \in a: F(x)\}$) and the scheme of substitution. It is very easy, however, to construct families of sets N in which the predicative scheme of set construction is valid. This scheme states the existence of the set $\{x \in a: F^a(x)\}$ where F^a is obtained from F by restricting all quantifiers to a. It is known that this restricted schema is satisfied in a family N provided that N is closed with respect to a finite number of operations, e.g., the eight well-known operations used by Gödel [9] in his definition of constructible sets. We shall call a family N predicatively closed if it is closed under these operations.

Thus the simplest way to obtain an extension of a given model M in which the predicative scheme of set construction will be valid is this: we add to M a new set a not yet contained in M and close it with respect to the eight operations. If M is denumerable, then a can be found already among sets of integers since their number is greater than the cardinal number of M.

More precisely the elements of the new family (which we call N or $M(a)$ in order to indicate its dependence on a) can be represented in the form $F_\alpha(a)$ where $F_\alpha(a)$ is the set obtained from a in exactly the same way as Gödel's F_α was obtained from 0. While the value of a is not fixed we can think of $F_\alpha(a)$ as of a polynomial of a kind with one variable a.

We have now to decide upon the choice of a in such a way that not only the predicative set existence scheme but also the unrestricted one, along with all the other set-theoretical axioms, are valid in $M(a)$.

The device invented by Cohen in order to obtain this result was a reduction of properties of N to the properties of M. Such a reduction is trivially possible when a is an element of M but of course this does not help us to solve the problem of independence. When a is completely arbitrary then the required reduction is clearly impossible, although I do not know how far the properties of $M(a)$ are independent of those of M. Sets a which are not in M but for which the reduction is still possible were called *generic* by Cohen. They can be defined relatively simply in terms of general topology. The definition given below is due to Ryll–Nardzewski.

Let us call two sets a, b of integers n-close if $i \in a \equiv i \in b$ for all $i < n$. The greater is n, the closer are the two sets. Sets which for an integer n are n-close to a form a neighbourhood of a. A family A of sets a is dense if in every neighbourhood of an arbitrary set there is at least one element of A.

Let now H be a formula with the free variables x, y, ... A dense family of sets a is called a generic family for this formula and its elements are called generic sets for this formula, if for arbitrary polynomials $F_\alpha(a)$, $F_\beta(a)$, ... the function $v_H(a) =$ *value of* $H(F_\alpha(a), F_\beta(a), \ldots)$ *in the model* $M(a)$ is a continuous function in A. Continuity means of course that for each a in A we can find an integer n such that v_H is constant for all arguments which are n-close to a and belong to A.

A family A is generic (and its elements are generic sets) if it is dense and generic for an arbitrary formula H.

Cohen proved the existence of generic families by introducing a new metamathematical notion of forcing. It has been noted by Ryll–Nardzewski that one can obtain the same result using a theorem of Baire which is one of the well known results of the descriptive theory of sets; see e.g. [11]. The function $v_H(a)$ is a Baire function (in case M is denumerable) and hence according to the Baire theorem is continuous on a residual set A. Thus the existence of generic families is established.

Let a be an arbitrary generic set and Π the family of its neighbourhoods. We can then reduce the properties of $M(a)$ to those of M in the following way. To each formula $H(x, y, \ldots)$ we let correspond a formula $\Phi_H(\pi, \alpha, \beta, \ldots)$ with the property

$$(*) \qquad \models_{M(a)} H[F_\alpha(a), F_\beta(a), \ldots] \equiv (\mathsf{E}\pi)_{\pi \in \Pi} \models_M \Phi_H[\pi, \alpha, \beta, \ldots].$$

The way to obtain this formula has been described by Cohen [5]; in his terminology the formula says that π forces the formula [1] $\neg\neg H(F_\alpha(a), F_\beta(a), \ldots)$.

On the basis of the equivalence (*) it is an easy matter to prove that all axioms of Zermelo–Fraenkel are valid in $M(a)$. The verification is particularly easy when one assumes (as one is entitled to do) that M is a union of an increasing well-ordered sequence M_i of its elementary sub-models satisfying the condition $M_i \in M_j$ for $i < j$.

We will not enter into the details of this verification which proceeds as in Cohen's and Solovay's papers [5], [21].

Cohen [5] and several other authors have successfully applied the method of forcing to various problems of independence. The number of such results increases at a disquieting rate. Most striking are results concerning the continuum hypothesis. Roughly speaking they show that practically every hypothesis concerning powers of regular cardinals is compatible with axioms of Zermelo–Fraenkel. Let me quote the following precise result due to Easton [6]. Let M be a denumerable model of Gödel–Bernays axioms and G an increasing function from ordinals to cardinals in M satisfying the condition that the sentence '$\aleph_{G(\alpha)}$ is not co-final with any cardinal $\leqslant \aleph_\alpha$' is true in M. Under these hypotheses there is a model N containing M such that both models have the same cardinals and the sentence $2^{\aleph_\alpha} = \aleph_{G(\alpha)}$ is true in N for every regular cardinal $\aleph_\alpha$.

Such results show that axiomatic set-theory is hopelessly incomplete. Certainly nobody expected the axioms of set-theory to be complete, but it is also certain that nobody expected them to be incomplete to such a degree. Since independence proofs are ob-

---

[1] Work is in progress towards a different definition of the formula Φ_H which avoids the notion of forcing and refers directly to the topological definition of generic sets.

tained by means of well-founded models there can be no hope that completeness will be restored by the use of some infinitistic rules of proof such as e.g. the rule ω. This shows that the incompleteness of set-theory is caused by other circumstances than the incompleteness of arithmetic. It is comparable rather to the incompleteness of group theory or of similar algebraic theories. These theories are incomplete because we formulated their axioms with the intention that they admit many non-isomorphic models. In case of set-theory we did not have this intention but the results are just the same.

Models constructed by Gödel and Cohen are important not only for the purely formal reasons that they enable us to obtain independence proofs, but also because they show us various possibilities which are open to us when we want to make more precise the intuitions underlying the notion of a set. Owing to Gödel's work we have a perfectly clear intuition of a set which is predicatively defined by means of a transfinite predicative process. No such clear interpretation has as yet emerged from Cohen's models because we possess as yet no intuition of generic sets; we only understand the relative notion of a set which is generic with respect to a given model. Probably we shall have in the future essentially different intuitive notions of sets just as we have different notions of space, and will base our discussions of sets on axioms which correspond to the kind of sets which we want to study. Although nothing certain can be predicted, we presume that there will be a common part of these various axiomatic systems, and that axioms belonging to this common part will describe the most primitive parts of set-theory which are needed in the expositions of mathematical theories perhaps including the category theory. I do not dare to speculate how these various systems of different set theories will decide the question whether sets of very high power exist.

In my opinion everything in the recent work on foundations of set theory points towards the situation which I just described. Of course if there are a multitude of set-theories then none of them can claim the central place in mathematics. Only their common part could claim such a position; but it is debatable whether this common part will contain all the axioms needed for a reduction of mathematics to set-theory.

It is not impossible that Cantor himself realized that his notion of sets was vague and capable of several interpretations. His curious remark on sets alluded to at the beginning of this lecture is not inconsistent with this possibility. At any rate, since the existence of these various interpretations is confirmed by recent developments, we must conclude, strange as it may seem, that Cantor's views on sets were probably nearer the truth than the seemingly clear intuitions of Dedekind.

References

[1] J. W. ADDISON, 'Some consequences of the axiom of constructibility', *Fundamenta Mathematicae* **46** (1958), pp. 337–357.

[2] O. BECKER, *Grundlagen der Mathematik in geschichtlicher Entwickelung*, München 1954, p. 316.

[3] P. BERNAYS, 'Zur Frage der Unendlichkeitsschemata in der axiomatischen Mengenlehre', *Essays in Foundations of Mathematics Dedicated to Prof. A. A. Fraenkel*, Jerusalem 1961, pp. 3–49.

[4] L. BUKOVSKÝ and K. PŘÍKRÝ, 'Some metamathematical properties of measurable cardinals', *Bulletin de l'Académie Polonaise des Sciences*, séries des sciences mathématiques, astronomiques et physiques. 14 1966, pp. 9–14.

[5] P. J. COHEN, 'The independence of the continuum hypothesis', Part I, *Proceedings of the National Academy of Sciences of the U.S.A.* **50** (1963), pp 1143–1148; Part II, *Ibid.*, **51** (1964), pp. 105–110.

[6] W. B. EASTON, *Powers of regular cardinals*, Unpublished doctoral dissertation, Princeton 1964.

[7] A. A. FRAENKEL and Y. BAR-HILLEL, *Foundations of set theory*, Amsterdam 1958.

[8] H. GAIFMAN, *Further consequences of the existence of measurable cardinals*. Forthcoming.

[9] K. GÖDEL, *The consistency of the axiom of choice and of the generalized continuum hypothesis with the axioms of set theory*, Princeton 1940.

[10] H. J. KEISLER and A. TARSKI, 'From accessible to inaccessible cardinals', *Fundamenta Mathematicae* **53** (1964), pp. 225–308.

[11] K. KURATOWSKI, *Topologie I*, Warszawa–Wrocław 1948.

[12] A. LÉVY, 'Axiom schemata of strong infinity in axiomatic set theory', *Pacific Journal of Mathematics* **10** (1960), pp. 223–238.

[13] N. LUSIN, *Leçons sur les ensembles analytiques et leurs applications*, Paris 1930.

[14] P. Mahlo, 'Zur Theorie und Anwendung der ϱ_0 Zahlen', *Berichte über die Verhandlung der Sächsischen Akademie der Wissenschaften zu Leipzig, Mathematisch Naturwissenschaftliche Klasse*, **64** (1912), pp. 108–112; **65** (1913), pp. 268–282.

[15] R. Montague and R. L. Vaught, 'Natural models of set theories', *Fundamenta Mathematicae* **47** (1959), pp. 219–242.

[16] A. Mostowski, 'An undecidable arithmetical statement', *Fundamenta Mathematicae* **36** (1949), pp. 143–164.

[17] P. S. Novikov, 'O néprotivoréčivosti nékotorych položenij déskriptivnoj téorii množestv, *Trudy Matématičeskogo Instituta im. V. A. Stéklova* **28** (1954), pp. 279–316.

[18] K. Příkrý, 'The consistency of the continuum hypothesis for the first measurable cardinal', *Bulletin de l'Académie Polonaise des Sciences; série des sciences mathématiques, astronomiques et physiques* **13** (1965), pp. 193–197.

[19] F. Rowbottom, *Large cardinals and small constructible sets*, Unpublished Thesis, University of Wisconsin 1964.

[20] D. Scott, 'Measurable cardinals and constructible sets', *Bulletin de l'Académie Polonaise des Sciences, série des sciences mathématiques, astronomiques et physiques*, **9** (1961), pp. 521–524.

[21] R. M. Solovay, *Measurable cardinals and the continuum hypothesis*, Mimeographed, Institute for Advanced Study, Princeton 1964.

[22] A. Tarski, 'Über additive und multiplikative Mengenkörper und Mengenfunktionen', *Comptes Rendus des Séances de la Société des Sciences et des Lettres de Varsovie, class III*, **30** (1937), pp. 151–181.

[23] A. Tarski, 'Über unerreichbare Kardinalzahlen', *Fundamenta Mathematicae* **30** (1938), pp. 68–89.

[24] A. Tarski, 'Some problems and results relevant to the foundations of set theory', *Logic, Methodology and Philosophy of Science. Proceedings of the* 1960 *International Congress. Stanford* (1962), pp. 125–136.

[25] S. Ulam, 'Zur Masstheorie in der allgemeinen Mengenlehre', *Fundamenta Mathematicae* **16** (1930), pp. 140–150.

Note added in proof. The construction mentioned in the footnote on p. 93 has been meanwhile completed. The topological version of Cohen's theory has been also found and carried out by G. Takeuti in a paper 'Topological spaces and forcing' (not yet published).

DISCUSSION

G. Kreisel: *Comments.*

First of all I wish to say how much I like the orderly and succinct presentation of a remarkable number of recent results including, in passing, several results not previously published. But I am equally struck by omissions of a number of simple basic facts. These facts are, I believe, philosophically at least as important as the mathematically much more interesting recent constructions stressed by Mostowski. These simple facts may well be more important for a fruitful choice of notions and problems: so, in the long run, as often happens, philosophical importance and *future* mathematics go hand in hand.

A. *Historically*, the principal omissions seem to me to be these.

1. The formal independence results confirm what has been morally certain for a long time, at least 20 years. Mostowski quotes Gödel's remarks on the need for a more precise analysis of the basic notions of set theory, presumably from Gödel's articles on Russell's logic and on Cantor's continuum problem [1]. But Mostowski seems to have read these articles very casually. The latter essentially takes the formal independence of the continuum hypothesis (CH) for granted, and discusses principles of evidence that may be needed. It is simply false (p. 89) that Gödel believed the *intuitive* notion of set to be too vague to decide the axiom of choice AC

[1] It is true that in Gödel's original note in the *Proc. National Acad.*, 1938, he remarks that $V = L$ added as a new axiom seems to give a natural completion of the axioms of set theory, in so far as it determines the vague notion of an arbitrary infinite set in a definite way. This remark is certainly not in the spirit of the article on the continuum problem in the *American Math. Monthly*, 1947, and even less in the spirit of the addendum to this article in the anthology by Benacerraf–Putnam, 1964. It is a pity that Gödel himself does not draw explicit attention to the change in his point of view. But it seems reasonable to attach more philosophical weight to a full discussion (1964, i.e. more than 25 years after the original research announcement) than to a single remark; particularly as the 1947 and 1964 publications develop the *same* point of view.

or the CH: in the article on Cantor's continuum problem he explicitly states that the AC has the same evidence for the intuitive notion of set (in the cumulative hierarchy of Zermelo) as the other accepted axioms. (And, personally, I find AC more *immediately* evident for this notion than the replacement axiom which one has to analyse first). It is trivially false (p. 83) that the axioms of Zermelo and Fraenkel were accepted from the beginning (because otherwise they would not have special names). Zermelo's notion is very appropriately called the notion of: set of something (Gödel's article on the continuum problem). The omission of Zermelo's own explanation of the meaning of his axioms is very bad: the explanations shows how the axioms are discovered and, from the explanation, it is clear that this process of discovery can certainly be continued, for instance, by use of the first principle on p. 84 of Mostowski's paper. (Even this appears *ad hoc* because he says next to nothing about the actual process of discovery which was successful in the past. Incidentally, it would be interesting to discover the logical notions needed to formulate this principle generally and precisely.)

2. It is perhaps interesting that Mostowski finds the implications of the existence of large cardinals for projective set theory very surprising. But it is also interesting, and historically more to the point, that Gödel, in the papers above (and also in his Princeton Lecture, 1948, recently published in Davis' anthology) discussed this possibility in detail. His discussion is not based on very sophisticated considerations, but on such well established results as the role of axioms of infinity for formal proofs of consistency, (e.g. of the existence of a set of rank $(\omega+\omega+1)$ for establishing the existence of models or well founded models of Zermelo's set theory). To this might be added the use of analytic number theory (sets of rank $\omega+1$ for complex numbers, of rank $\omega+2$ for functions of them) to decide arithmetic statements (about sets of rank ω). If assumptions about sets of higher rank lead to *simpler* proofs of known results, at least *prima facie* the presumption is that they lead to new results, given that the known (recursively axiomatised) theory of hereditarily finite ($=$rank ω) sets is incomplete. (It is an important *discovery* of delicate proof theory that in analytic number theory so far developed the existential as-

sumptions mentioned can be eliminated.) Of course it would be stupid to say that the *truth* of theorems about sets of low rank depends on the truth of axioms of infinity, but their *evidence* may do; just as in the case of numerical arithmetic a complicated statement may only be evident via general considerations (though the truth of numerical statements happens to be *recursively* decidable, in contrast to the truth of statements about sets of rank α, $\alpha \geqslant \omega$).

3. It seems to me wrong not to take seriously the very real philosophical problems which presented themselves to people before they were solved by Zermelo. Mostowski mentions predicativity: it is a fact that 70 years ago most set theoreticians and mathematicians were, in practice, predicatively inclined. Axioms of set theory which they formulated were weak. *The axiom of reducibility looked at least as unjustified as the measurable cardinals.* The people failed to find a justification, not because they all *wanted* predicative axioms in the strict sense [cf. Feferman, *Journal of Symbolic Logic* **29**, 1964, pp. 1–30, when reducibility is not only false, but even its consistency is not predicatively provable], but because they saw no way of getting beyond predicative axioms. Zermelo's set existence (comprehension) axiom is formally similar to the axiom of reducibility. He justified it, not by formal relative consistency proofs (p. 87), but by explaining, in words, the notion of: set *of* something.

B. *Details.* 1. Mostowski rather dismisses the proofs of independence of axioms of infinity from the usual axioms ZF as simple (p. 87) in contrast to independence of CH (p. 90). He does not emphasize that the simple proofs establish something much stronger, namely (cf. p. 100) independence from the (strong) second order form of ZF. (For more detail, cf. my paper in this volume, § 1 (b), and the difference between (i) independence of the parallel axiom in geometry and (ii) of the CH in set theory). – Incidentally I bet that people who have not explicitly acquired Zermelo's notion would not find for themselves these 'simple' independence proofs.

2. Mostowski emphasises (p. 94) that CH is not decided by the ω-rule in contrast presumably to Gödel's undecidable arithmetic statements. He fails to say that CH *is* decided by the second order axioms of Zermelo (in contrast, here, to the axioms of infinity just

mentioned). It would certainly be interesting to analyse seriously the assertion (p. 94) that 'the incompleteness of set-theory is caused by other circumstances than the incompleteness of arithmetic'. I conjecture that a serious analysis would distinguish between (i) the second order independence of axioms of infinity (how *often* the collecting process is to be repeated) and (ii) the independence of CH. Of course, there is a distinction between independence of arithmetical statements and of the CH in terms of ω-models and so-called β-models, but this is purely technical, at least without further analysis.

3. Mostowski finds the number of independence results disquieting (p. 93). Does this show more than the cleverness of mathematicians in formal manipulations? The set-theoretic questions which they have failed to decide are formally undecided by the axioms so far recognised. Is this 'large' number of independence results not simply parallel to the following familiar fact? When a new axiom or a new method is needed and/or discovered a whole batch of questions that interest people is affected.

C. *Conclusion.* Mostowski's main conclusion that only the common part of a multitude of set theories can claim the (a?) central place in mathematics (p. 94), seems a perfect non-sequitur. Even granted different intuitive notions of set (e.g. predicative and Zermelo's) there is still the question which is fundamental, and therefore central. If concept X can be defined in terms of Y but not conversely, then Y is more fundamental. Examples: (a) The notion of constructible set (essentially: of ordinal and certain 'recursive' functions of ordinals) can be defined in terms of Zermelo's notion, but not conversely. (b) Sets of rank $<\omega+\omega$ can be defined in terms of sets of rank $<$ first measurable cardinal, but not conversely. Nothing in his paper excludes the possibility that there is such a thing as a fundamental theory.

I realise of course that I have made no positive suggestions for deciding the open questions. Since the discovery of axioms is, for me, the most interesting side of foundations I have personally concentrated more on constructive (non-set-theoretic) logic where one can rely on a non-trivial new axiom a year because the subject is less explored.

I think the gist of Mostowski's reply to what I just said would be that second order axiomatisations have no place in an analysis of the notion of 'set' because such axiomatisations assume that we already know the notion of set. But surely the underlying (and incidentally, quite common) assumption in this is that foundational analysis should take the form of a reduction; for instance, to a first order formalisation or (for reasons which could be made more explicit though he has not done so) to a formalisation in terms of a consequence relation defined by means of a generalised inductive definition, such as the ω-rule. Probably, though this is not stated, there is the further assumption that once the reduction is given one 'loses nothing' by forgetting the 'abstract nonsense' concerning the meaning of the notions analysed: one can take the formalisation itself as the new starting point. For example, because of this one has ignored Zermelo's derivation of his axioms from the notion: *set of*, even when one studied the formal properties of these axioms.

Though I discuss the defects of the assumptions implicit in Mostowski's reply (and paper) in my own contribution in general terms, this is to be said here:

First, philosophically, the assumption is incoherent if one wishes to interpret mathematical notions set-theoretically at all. The notions of first order and second order consequence are defined in terms of the *same* concepts; only from a formal point of view is there a difference: the main conclusions are derived from different properties of the same basic concepts; e.g. different instances of the comprehension axiom are used to show the basic facts that (i) the set of valid first order sentences is recursively enumerable and (ii) the set of valid second order sentences is not. I do not deny that there is an essential philosophical distinction between first order and second order reasoning for a non-semantic, e.g. for a finitistic, foundation à la Hilbert, but it is bad philosophy to transfer thoughtlessly a distinction from one foundational framework to another.

Second, quite naively, it is simply not true that second order notions are 'unclear'. Not only were the first axiomatic theories, e.g. of Dedekind, intended to be categorical, and, as we now say, second order: i.e. they appeared first philogenetically. But one also learns more readily the idea of *arbitrary* set-of, e.g. set of numbers, than, say, the notion of: recursive or arithmetic set. All mathematicians know the former, most get bored when they are supposed to learn the latter.

Philosophical clarification of a concept does not usually consist in reduction to another set of concepts (this is the assumption in question) but consists in becoming aware of the properties of the concept; so to speak, looking at the concept, and not looking past it. This is what Zermelo did; people seem to have forgotten how to do it. As I said in my comments it is easier to do this when one studies a subject that not too many people have looked at, e.g. fundamental notions of constructive mathematics. The fact that this can be done at all shows that the further assumption (forget about 'the meaning of the notions') is even mathematically bad.

Specific example of fruitfulness of second order approach: If one takes it seriously at all, one will seek to formulate the comprehension and replacement axioms for *all* properties; or, at least, if only for definable properties, one will try to widen as far as possible the definitions considered (i.e. beyond first order language with $\in$ as sole constant). Thinking of infinite formulae one is led to find finite formulae (of higher type) which express the meaning of these infinite formulae: this is a principle for getting new axioms, not mentioned by Mostowski. In the present state of affairs this principle is not very practical because (as far as I know) all axioms got in this way can also be got by quite simple reflection principles, but not conversely. In other words, the latter are more fundamental. But the existence of this principle shows clearly that taking second order notions seriously has at least some consequences for the discovery of new axioms.

Added in print. Early doubts about the axiom of choice almost certainly depended on interpreting, perhaps unconsciously, the variables (in set-theoretic statements) as ranging over *definable* or *namable* sets, the notion of absolute definability being tacitly understood. If this notion is accepted at all then the comprehension axiom is evidently valid, but the axiom of choice quite dubious.

(If the variables are taken to range over sets that are 'first-order-definable-from-ordinals', i.e. Gödel's constructible sets, the axiom of choice is evident and, less obviously, the comprehension axiom is valid.) It should be noted that the consequences of the axiom of choice which are usually found to be paradoxical, affect principally the identification of the notions of (i) *geometric figure* and (ii) *arbitrary set of points*, e.g. Hausdorff's sphere paradox depends ultimately on explicitly defined denumerable disjoint sets A, B, C such that A can be rotated both into B and C separately, and also into $B \cup C$. This is 'geometrically' paradoxical, but nothing else but an elaboration of $\aleph_0 + \aleph_0 = \aleph_0$, and so has nothing to do with the axiom of choice.

A. Robinson: *Comments.*

Nobody would wish to deny the great interest of much of the work on independence and consistency in Axiomatic Set Theory, which has in recent years achieved a new high. There remains the problem of gauging the significance of this work for the foundations of Mathematics. Are the results obtained so far provisional in the sense that we may expect the axioms of Zermelo–Fraenkel to be supplemented in due course by new axioms whose truth would seem to be self-evident and which would enable us to decide e.g. the continuum hypothesis? Put in this form the problem is a matter of historical prediction concerning the beliefs of a certain social group at some future date. But many of those present would no doubt like to go further and would maintain that the wide acceptance achieved by the axioms of Zermelo–Fraenkel is merely a manifestation of their basic validity and that the same should apply to new axioms, yet to be discovered. On the other hand those who, like myself, have a deep distrust of ontological assertions concerning infinite totalities will find some support for their attitude in the current bifurcation (or multifurcation) of Set Theory and will even consider the possibility of a similar bifurcation in Number Theory. This is not to say that I would welcome the spreading of total anarchy in the foundations of Mathematics, nor even that I consider such a development likely. Indeed, as I have explained elsewhere (at the Jerusalem Congress, 1964), while I would describe myself as a formalist, I feel compelled to admit the existence of certain basic forms of thought – in Logic, in Arithmetic, and perhaps in Set Theory – which are prior to the arbitrary choice of mathematical axioms. Perhaps one may hope that out of the present difficulties there will arise a deeper understanding of this common core of all Mathematics.

L. KALMÁR: *On the role of second order theories.*

First, let me comment on *the role of second order theories.* While Professor Mostowski used second order logic in order to construct arithmetical sentences which are undecidable in some axiomatic set theories. I have the impression that for Kreisel second order methods are as useful and as correct as first order ones. I do not agree with him. I mean, the Neumann–Bernays–Gödel axiomatisation of set theory was a great success because it eliminated second order logic (as well as the need for an infinity of axioms) from set theory. It turned out that first order logic is far more convenient for proving a set theoretical theorem (e.g. for all ordinals) than second order logic (or axiom schemes). Well, second order logic may help us to gain meta-theorems which are different from, or even contradictory to those which we have got by first order logic, e.g. in the categoricity problem. However, I think, second order categoricity results are *deceiving*: they serve only to puzzle ordinary mathematicians who do not know enough logic to distinguish between first order and second order methods. One can say humorously, while first order reasonings are convenient for proving true mathematical theorems, second order reasonings are convenient for proving false metamathematical theorems. Of course, instead of calling them false we honour them by calling them second order meta-theorems. However, I do not think second order categoricity theorems can serve any sound purpose. This, of course, does not exclude the use of second order methods for establishing interesting undecidability or independence results.

As to the question why we should accept such strong axioms of infinity in set theory, my opinion is that we are ready to accept them because we want to regain as much of the 'lost paradise' of the naive set theory as possible. But I agree with Mostowski that none of the possible set theories can claim the central place in mathematics which Cantor's set theory would hold if there were no paradoxes. I guess that in the future we shall say as naturally 'let us take a set theory S' as we take now a group G or a field F. Of course sometimes we shall take a set theory S so and so (i.e. an arbitrary model of some axiom system of set theory).

A. Mostowski: *Reply.*

There are two general questions which occur in all the three contributions of the participants in this discussion.

One of them is the problem of the role of axioms of set theory. Are the axioms invented by Zermelo and Fraenkel (and other similar axioms) in any sense privileged and if not what are new axioms which should be added to ZF or which should replace these axioms?

The second problem is the possible role of higher order logic in the foundations of set theory. Besides these two problems Professor Kreisel voices a severe criticism of my paper and makes a number of interesting observations.

Let me treat first the two general questions. I quite agree with Professor Robinson that the importance of the recent metamathematical discoveries in set theory depends in the last analysis on the role of the Zermelo–Fraenkel axioms. He envisages a bifurcation or even a multifurcation of set theory. As I said in my paper I completely agree that this is the most probable course the abstract theory of sets will take in the future. The picture which Professor Kalmár describes in the second part of his remarks is essentially not different from the one which Professor Robinson considers. I can only add that various set theories which will perhaps appear in future must be based on a firm intuitive basis. Otherwise it is hard to see what would be their role. Thus it is not at all clear that one will have such a degree of freedom in the choice of a set theory as one has at present in, say, group theory.

If we assume that the various rival set theories will have the same intuitive appeal, then I think the conclusion is inescapable that none of them can be taken as a central mathematical theory. In my opinion our inability to decide the choice of set-theoretical axioms is an indication that no best set theory exists and thus I cannot agree with the final statement in Professor Kreisel's remarks (first paragraph of his 'Conclusion').

When one reads the text of Professor Kreisel's remarks one has the impression that he considers the notion of a set as a collection of objects ('set of') as perfectly clear. His statement about the intuitive evidence of the axiom of choice seems to me rather bold in view of the paradoxical sets whose existence is provable with the help of this axiom. We must also remember that the majority of so-called working mathematicians always accepted the intuitive notion of a set as a 'collection' and yet were not able to decide a multitude of seemingly very simple questions. One is thus entitled to wonder whether the notion of an arbitrary collection of objects is as simple and as clear as it seems at first sight.

Of course it is not impossible that no essential progress was made in the study of, say, the generalized continuum hypothesis simply because the problem is too difficult to solve by two or three generations of mathematicians.

In view of the enormous success which Gödel achieved with his notion of constructible sets in the study of various set-theoretical problems the idea (also formulated by Gödel) that the vague notion of 'sets' may be in need of clarification gains support. On the other hand no reasons compel us to limit ourselves only to constructible sets and Cohen's theory supports us in the reluctance to admit this axiom. One is thus led to accept the idea of a bifurcation or even a multifurcation of set theory. This picture has to me a much stronger appeal and seems to correspond better to the actual state of set theory than the belief that the notion of an arbitrary collection of objects is clear.

I do not want to discuss here at length the problem of logic of second order and its application to foundation of set theory because I said what I think about these matters in my contribution to the discussion of Professor Bernays' lecture (pp. 113–114). In view of what Professor Kalmár said in the first part of his remarks I want to say only that in my opinion logic of second order is a part of set theory. Thus it seems to me that it is a perfectly sound part of mathematics and that its theorems, as well as metamathematical theorems proved for systems based on this logic, are neither misleading nor false. They just require stronger assumptions than many metamathematical theorems concerning systems based on first order logic. It is true, as Professor Kreisel points out, that some metamathematical theorems concerning systems based on first order logic also require an amount of set theory; thus in this respect there is no essential difference between theories based on first and on second order logic.

The situation becomes different when we try to apply second order logic to the foundation of set theory. We agreed that we must look for new set-theoretical axioms because the old ones proved insufficient. Second order logic gives us immediately an easy answer: one can for instance accept the following definition: x is a set if and only if it is either void or

$$(\mathrm{E}t, z)\{(x \in z) \wedge (t \in x) \wedge (\mathrm{E}F)(y)[y \in x \equiv F(y)]\}.$$

All problems of set theory are then reduced to problems of second order logic. Can we expect anything from such an approach? Professor Kreisel is right that the continuum hypothesis is decided in second order logic but neither he nor anybody else can tell us whether it is decided in a positive or in a negative sense. In order to decide this question one would have to decide the continuum hypothesis. Thus Professor Kreisel has simply reformulated the problem. To sum up: second order logic is a form of set theory and we cannot expect that we will gain much by considering both theories simultaneously.

Perhaps this is bad philosophy but I think that it seldom happens that one clarifies a notion by discussing along with it another one that is exactly as involved and complicated as the former.

Concerning specific questions and comments of Professor Kreisel:

(i) The view that the notion of set is vague and that one can make it more precise was expressed by Gödel in his paper: 'The consistency of the axiom of choice and of the generalized continuum hypothesis', *Proc. Nat. Acad. Sci. US* **24**, 1938, pp. 556 f.

But Kreisel is right in stating that Gödel did not maintain this view in his (later) paper on the Continuum Hypothesis. I personally found the view expressed in the earlier paper of Gödel extremely illuminating and I regret that I did not at once quote the source.

(ii) I doubt whether most mathematicians who were active 70 years ago adhered to the point of view of the French school who certainly considered the predicative notion of a set as the only acceptable one. Cantor, Frege, Dedekind are clear counterexamples.

(iii) I share Kreisel's opinion that I failed to give in my paper a comparison of the results of Gödel concerning the role of the axioms of infinity in proofs of arithmetical sentences and of the role of strong axioms of infinity in deciding e.g. the axiom of constructibility. All I can say is that I feel that there is a deep difference between these results. Undecidable propositions constructed by Gödel and then decided by means of axioms of infinity were called by Gödel himself formally undecidable, whereas nobody would call e.g. the continuum hypothesis an only formally undecidable statement. But I agree with Kreisel that this distinction is vague and that the problem should be analyzed further.

(iv) Concerning the question whether Zermelo–Fraenkel axioms were at once accepted by the working mathematicians I would be inclined to believe that those mathematicians who were working on abstract set theory certainly did so. Of course the representatives of the intuitionistic or semi-intuitionistic schools did not accept the axioms, which is only understandable if one reflects that they dismissed abstract set theory altogether.

(v) My teaching experience shows that the independence of the axiom of infinity from the remaining axioms of ZF (whether based on first or on second order logic) *is* a trivial result. I confess that I fail to see why Professor Kreisel thinks it is not.

WHAT DO SOME RECENT RESULTS IN SET THEORY SUGGEST?

PAUL BERNAYS

Zurich

The considerations I want to present in this paper are closely related to the problems discussed by Professor Mostowski [1].

The first essential thing which emerges from his paper is that the results of Paul J. Cohen on the independence of the continuum hypothesis do not directly concern set theory itself, but rather the axiomatization of set theory; and not even Zermelo's original axiomatization, but a sharper axiomatization which allows of strict formalization.

It is wellknown that any strictly formal system of set theory is subject to the Skolem paradox: which means that the axioms are satisfiable in a denumerable model. Cohen makes essential use of this, and his reasonings even sharpen the Skolem paradox. At all events the models which he presents have the character of non-standard models.

In general the possibility of non-standard models of an axiom system is due to the presence of a principle in which a concept of set, or sequence, or predicate, occurs. In order that the model theory of an axiom system containing such a principle has a standard character, we have to identify the said concept with the corresponding concept of the model theory. This does not preclude formalization; the model theory of a theory T is then to be represented by a formalized set theory S, and when a principle of T contains a set concept, this has to be formalized by the set variable of S. In this way it is, for instance, possible to prove formally the categoricity of number theory.

When the theory T is itself axiomatic set theory, we have to

[1] See his paper 'Recent results in set theory'; this volume, pp. 82–96.

deal, when we consider models of T, with two set theories. Cohen's proof also proceeds, so to speak, by a combination of two set theories: the one by which the transfinite powers are distinguished, the other in which the denumerability of the original model is stated. But whereas the Cohen procedure leads to non-standard models, we can, by the forementioned device, state a kind of categoricity of the axioms of set theory, as was done, without formalization, by Zermelo in his 'Grenzzahlen und Mengenbereiche'[1].

Thus we see that the independence of the continuum hypothesis is essentially tied to the formalization of set theory. It is therefore a fact of a similar kind to the existence of non-standard models for formalized number theory.

We might now ask how far these facts seem to support the 'formalistic' doctrine. There are stronger and weaker forms of this doctrine.

According to an extreme form, the significance of a mathematical theorem consists merely in the fact that it is found to be provable in an adopted formal deductive system. This view is in any case defective. For instance, even when we can derive a formula 'for all x: $A(x)$' in the adopted system, we do not thereby know that $A(x)$ really holds for every x, unless a consistency proof is given for the adopted system. But this then is an *intuitive* proof of a general number-theoretic theorem which must be understood *in the normal way* in order to yield the wanted result. This point has been repeatedly stressed by Georg Kreisel.

A much more moderate formalistic doctrine says that proper mathematics deals with *constructive reasonings* about provability and unprovability; although in addition to directly constructive mathematics, formal derivations are also to be admitted, provided they are performed in systems which either are shown to be consistent or, in virtue of long experience, assumed to be consistent. This is at any rate a sensible doctrine. However, among the arithmetic propositions which have the form of *possible* constructive theorems, there are many which are as undecided as the classical continuum problem; and so by this more moderate formalistic

---

1 *Fund. Math.* XVI, 1930.

doctrine mathematics gets no less a fragmentary character than it does by the classical doctrine. Moreover, from our experiences with non-standard models, it appears that a mathematical theory, like number theory, cannot be fully represented by a formal system; and this is the case not only with regard to derivability, as Gödel's incompleteness theorem has shown, but already with regard to the means of expression. For a full representation we need an open succession of formal systems.

All this hardly supports a formalist view of mathematics. Nevertheless the proof-theoretic investigation, in the sense of a modified Hilbert programme, is at all events fruitful. In particular it is in the light of it that we can best appreciate the Cohen result about the independence of the continuum hypothesis: with regard to our set theoretic investigations this result is a disappointment; in model theory one might argue that the original question has not been answered, as Cohen's proof applies only to a *formalized* axiomatization of set theory. But for the *proof-theoretic* investigations the result has definitive significance, and here, of course, it amounts to a great success.

We have also to admit that, even if we transgress the domain of formalized proofs and adopt an *unrestricted* concept of a predicate, this does not yet provide us with methods for attacking the unsolved problems, and in particular the continuum problem in set theory.

Thus the situation can be described as follows:

(1) Proof-theoretically we have to acknowledge that there is an infinity of systems of Zermelo–Fraenkel set theory, any two differing as to the power of the continuum, and all being consistent provided the original Zermelo–Fraenkel set theory is consistent.

(2) If the strictly formal methods in axiomatic set theory are transgressed by applying the schema of the selection-axiom and that of the replacement axiom with an *unrestricted* concept of predicate, then model theory shows that the power of the continuum must be the same for each model of the Zermelo–Fraenkel axioms; yet we are not able to determine, by any of the known methods, what in fact is the power of the continuum.

Our inability to deal successfully with the continuum problem is certainly connected with the circumstance that our explicit knowledge of the continuum is very restricted. We are not even able to define effectively a subset of the continuum which can be shown to have the power of the second number class. So the theory of transfinite powers is fragmentary already in its beginnings.

Moreover, as we know from the antinomies, the transfinite powers do not constitute, as do the natural numbers, a mathematical structure. It therefore would be an illusion to think that we can exhaust the possible processes for generating powers.

No doubt, speculations about high powers have their great interest. But this task of set-theoreticians should be distinguished from the other task: to provide a suitable set-theoretic frame for the central mathematical disciplines.

The requirement of mathematical objectivity does not preclude a certain freedom in constructing our theories. We should use this freedom to build our mathematical theories in such a way as to be, as far as possible, comfortable homes for our intellect.

DISCUSSION

A. Mostowski: *Cohen's independence proof and second order formalisation.*

I want to comment on the distinction made by Professor Bernays in his article between the axiomatised and formalised set theories. According to Professor Bernays, Cohen's independence result is definitely established only for the latter. In this connection I want to stress the fact that there exist in the literature proofs of Cohen's results which do not assume the existence of denumerable models for set theory. Such a proof was given by P. Vopěnka in his paper, 'The Independence of the Continuum Hypothesis', *Comment. Math. Univ. Carolinae*, Prague 1964, the summary of which also appeared in the *Bulletin of the Polish Academy of Science*, vol. 13, 1965. Vopěnka's proofs proceed by interpretation. He defines in the language of the Gödel–Bernays set theory a new predicate $\in'$ which satisfies all the axioms and the negation of the continuum hypothesis. The predicate $\in'$ depends on a parameter j which denotes an ultrafilter in a suitable Boolean algebra. Thus Vopěnka proves in the Gödel–Bernays set theory the statement

$$(\mathsf{E}j)\{BG(\in') \wedge \neg H(\in')\}$$

where BG is the conjunction of the axioms, H is the continuum hypothesis expressed in the language of the Gödel–Bernays system and $BG(\in')$, $H(\in')$ are formulae obtained from BG and H by replacing everywhere the atomic formulae of the form $u \in v$ by $\in'(j, u, v)$.

In view of this result the statement that 'the independence of the continuum hypothesis is essentially tied to the formalization of set theory' does not appear to be justified. The usual axiomatisation is just sufficient provided it uses only finitely many first order axioms.

Many mathematicians not working on foundations are nowadays ready to accept the Gödel–Bernays system as *the* set theory. If this point of view is accepted then the conclusion seems inescapable that the continuum hypothesis is independent of the axioms in exactly the same sense as for instance the axiom $(x, y)(xy = yx)$ is independent of the axioms of group theory.

In view of what Professor Bernays said about the 'transgression of the axiomatic standpoint' I think that he would be inclined to take this result as an indication that the identification of the Gödel–Bernays system with the intuitive set theory is not justified. This at any rate is my point of view. We need new axioms to codify the intuitive set theory. The disquieting fact is that we do not know where to look for them.

Let me add still a few words about the second order axiomatisation of set theory. I quite purposefully refrained from comparing Cohen's result with the independence of Euclid's fifth axiom. The latter result holds for second-order axioms of geometry as well, whereas, as Bernays noticed, the Zermelo–Fraenkel axiomatic set theory based on second order logic has a kind of categoricity which is not possessed by the absolute geometry based on this logic. This point was also stressed in Kreisel's paper. However, as was pointed out in these two papers and especially in Bernays' paper, we obtain no additional information on the status of the continuum problem if we pass from the first order logic to the second order one.

This is evidently caused by the fact that second order logic with its semantic notion of consequence is (at least at present) not an operative system. We need axioms which will characterize the notion of an arbitrary predicate. The solution of the continuum problem depends essentially on the choice of these axioms. But is the problem of their choice in essence not the same as the problem of finding suitable axioms for the notion of a set? Sets considered in mathematics are by their very nature second order objects. It does not seem to me that anything essential can be gained by the duplication of the problem and by stating it separately for sets considered as elements of the universe of an axiomatic theory and for sets considered as 'extensional predicates'.

Y. BAR-HILLEL: *The dangers of Platonistic modes of speech.*

At the danger of repeating comments made on other talks presented here, let me insist once more that phrases like 'our explicit knowledge of the continuum is very restricted', used by Professor Bernays, are apt to mislead their users into believing that somewhere (perhaps in some Platonic heaven) there lies spread out the ideal continuum to be known and beheld by all people with sufficient uncluttered intuition and with sufficient readiness and training to use their mental experience. The requirement of mathematical objectivity must, and can, be met without recourse to such doubtful metaphors, to put it mildly.

As to Mostowski's comment, let me insist once again that I fail to find in myself something I would be tempted to call a conception of 'the intuitive set theory' and that I have grave doubts as to what other people mean exactly when they use this phrase, in particular with the definite article. While Mostowski finds it disquieting that he and many other mathematicians do not know where to look for the new axioms that would finally codify the theory, I for one rather find Mostowski's uneasiness disquieting.

P. SUPPES: *After set theory, what?*

As the papers of Bernays and Kreisel have emphasized, second-order formulations of set theory offer promise and should be thoroughly explored. All the same, it seems to me that Mostowski is quite right in his comments on Bernays' paper. In one clear sense Cohen's independence results have a standard and natural interpretation that settles decisively the independence of the continuum hypothesis without restriction to a particular formalization. Unless some real surprises come out of second-order set theories, it would seem that the conceptual situation in set theory is now very similar to that in geometry after the proof of the independence of the parallel postulate. Standard set theory, for instance, the Gödel–Bernays version, will continue, like Euclidean geometry, to occupy an important place in mathematics, but the search for a new sort of foundation will be inevitable. It seems unlikely that we shall be content to accept the view that there are many possible distinct foundations of mathematics because there are many possible set theories. Once again the search will be underway for a single unifying formulation of the foundation. It is certainly too early to forecast with any accuracy the main direction in which this search will move, apart perhaps from predicting a much intensified investigation of second-order logic.

There is, however, one possible direction that I should like to call attention to and indicate why I think it may be promising. For reasons that are not difficult to understand the classical emphasis in foundational research has been on the nature and construction of mathematical objects. The intensive development of model theory over the past thirty years has accentuated this interest. On the other hand, in both formalism and intuitionism there are strong tendencies that could easily lead to a concentration on the nature of mathematical thinking. Kreisel's recent work on finitist proof-theory and informal rigor provides some concrete examples. The burgeoning science of computers contains others. Questions about the nature of mathematical objects seem almost out of place in the context of computer programming, including theoretical studies on the characteristics of particular languages like ALGOL or FORTRAN. I can think of no deeper question in the philosophy of mathematics than to ask for a detailed specification of the programming language of the brain. The fact is that we as yet know very little about how we do think about mathematics. I realize that a question of this sort seems far removed from the main issues directly raised by Bernays' paper. The relevance lies in the attempt to forecast a direction of foundational research that will be more or less orthogonal to the search for a new and better, perhaps second-order, set theory.

P. Bernays: *Reply*.

I much appreciate the comments on my paper presented by Professors Mostowski, Bar–Hillel and Suppes. They give me an opportunity to add some clarifying and supplementary remarks.

Some criticism has arisen, I believe, from a misunderstanding caused by my abbreviated manner of expression: at the beginning I distinctly spoke – in comparison with Zermelo's original axiomatization of set theory – of 'a sharper axiomatization which allows of strict formalization'; later on I spoke simply of 'formalization of set theory' and of 'formalized axiomatization of set theory', but this was meant in the same sense. For the applicability of Cohen's result it does not, of course, matter – and here I fully agree with Mostowski and Suppes – whether axiomatic set theory is presented as a formal system or not, and even less in what particular system it is formalized. What the applicability does depend on is the mentioned sharper axiomatization by which strict formalization becomes possible, that is, the Fraenkel–Skolem delimitation of Zermelo's concept '*definite Eigenschaft*'.

Professor Bar-Hillel's criticism is likewise directed against a form of expression in my paper. I spoke of our 'knowledge of the continuum' in the usual way we do in mathematics, where the continuum is accepted as an object of investigation, like, for instance, any topological group or the Hilbert space. Bar-Hillel is opposed to this way of speaking; he wants to remind us of the possibility of a more reserved attitude towards mathematical objects. Such a reserved attitude is indeed at the basis of the task of proof theory. I did, however, not neglect in my paper the proof-theoretic aspect; I even remarked that it is under this aspect that Cohen's result can be best appreciated.

Professor Suppes mentions constructive proof theory as one of the actual tendencies which may lead to a concentration on the nature of mathematical thinking, and in particular in the direction of searching for 'a detailed specification of the programming language of the brain'. This task actually amounts to seeking suitable standard forms for formation of concepts and setting of problems. Combinatory logic tends to something like this. But perhaps it has not yet found sufficiently suitable patterns. If we should succeed in the said task, then perhaps set theory in its present form may become redundant and we may get rid of our present difficulties.

ON THE RELEVANCE OF POST-GÖDELIAN MATHEMATICS TO PHILOSOPHY

STEPHAN KÖRNER
Bristol University

Grounds of conflict between the results of a scientific discipline and any philosophy of it, be it analytical or programmatic, may range from real or apparent incompatibilities to implausible, or even merely uncomfortable, divergences. The philosophy of mathematics in so far as it analyses the structure of mathematical thought, may conflict with mathematics not only by being mistaken, but also by, for example, having lost touch with the subject or having been left behind by its actual development. In so far as the philosophy of mathematics suggests programmes for mathematics, these may turn out mathematically incapable of being executed, or they may simply be ignored and remain mere programmes. These features the philosophy of mathematics shares with the philosophy of any other science. From the philosophy of other disciplines it differs, however, in an important respect: while e.g. physical theories cannot in turn become the subject matter of physical theories, it is characteristic of mathematical theories that they can themselves become the subject matter of mathematical theories. It is thus in principle possible for mathematical theories and philosophical theories about mathematics to be incompatible. Although such conflicts, e.g. between mathematical and philosophical theses about geometry, have engaged the attention of philosophers for a considerable time, they have arisen in a more general, and therefore more acute, form since the discovery of the so-called limitative theorems by Skolem, Gödel, Tarski, Church and others and since the recent discovery of P. J. Cohen of the independence of the continuum hypothesis.

In this talk, after some remarks about the philosophical conceptions of mathematics up to and including the discovery of non-Euclidean geometries, I shall discuss some of the philosophical problems arising from this century's mathematical discoveries about mathematical theories, concluding with some tentative suggestions for their solution.

1. *On the philosophy of non-competitive mathematical theories.* Without begging too many questions one may, adopting and slightly adapting the convenient language of Leibniz, say that a proposition is true-or-false either in the actual though not in every possible world, or in every possible and therefore in the actual world. The actual world has, of course, been conceived in very different ways, e.g. as being or containing the world of sense-experience, a mind-independent world of Platonic Forms, or a partly mind-independent universe of Kantian phenomena. But the actual world is at least regarded as intersubjective. 'Propositions' which are 'true-or-false only in a possible world', such as the world of Dickens, are nevertheless capable of standing in logical relations to each other and of being, interestingly or uninterestingly, unified in deductive systems. This is one reason for defining propositions, not in terms of their having truth-values, but in terms of their capacity of standing in logical relations. Another is that, e.g., rules or norms of various kinds which, though they do not possess truth-values, logically imply other rules or norms, are commonly regarded as propositions. Lastly – and this is my main reason for accepting it – the wider definition leaves open a number of questions about so-called theoretical propositions in the natural sciences and about mathematical postulates and theorems, questions which are easily prejudged or overlooked if one employs the narrower definition of propositions as true-or-false, and the dichotomy of all true propositions into empirical and logically necessary ones.

All major philosophers from Plato to Kant and most later ones regard the postulates and theorems of mathematical theories as true in the actual world, however much they differ in their conception of it. Apart from Plato and Kant themselves, this applies also to Mill and very likely to many unknown pre-platonic empiricists who, like Mill, seem to have held that mathematical truths describe very general features of experience. Aristotle who, as it were, put the Platonic Forms into the physical objects, remained faithful to the doctrine of the truth of mathematical postulates and theorems by not admitting, or at least not considering, the possibility that conflicting mathematical theories can with equal correctness be abstracted from experience. The thesis of Leibniz and his logicist successors that mathematics is reducible to logic, also implies the doctrine of (one) mathematical truth.

This general doctrine of uniqueness is combined, and intimately connected, with other philosophical doctrines. It is thus generally agreed by these philosophers that every mathematical truth is in principle accessible to the human mind. This does not mean that every mathematical truth will in fact be discovered or that even every trained mathematician would be able to discover it, though he would be able to understand its proof.

It is, moreover, generally or almost generally assumed that it is in principle possible to find routine methods – applicable by every intelligent person – which will yield the answer to every mathematical problem. Plato had such hopes for his, or an improved, dialectic. Leibniz expressed them by his famous 'Let us calculate'. Kant might have expressed them by 'Let us construct the relevant concepts' and Mill by 'Let us apply the canons of induction with the utmost generality'.

A further common assumption made by all these philosophers concerns the relation between mathematical reality and thought or language. They all assume that it is possible for mathematical propositions or statements to reflect conceptually or linguistically their non-conceptual or non-linguistic subject-matter unambiguously and exhaustively. In this respect the subject-matter of mathematics differs, especially according to Plato and Kant, from sense-experience and aesthetic experience which always, as it were, over-flow any attempted reflection of them in propositions or linguistic formulations.

Until the discovery of non-Euclidean geometries the doctrines of the uniqueness of mathematical reality or intersubjective intuition, of its accessibility, of the solvability of all classes of mathematical problems and of the possibility of the unambiguous and exhaustive reflection of mathematical reality in conceptual or linguistic formulations were neither called into question nor regarded as needing a more precise analysis. Their questioning and analysis had to wait for the emergence of conflicts between mathematical theories and of conflicts between philosophical and mathematical theories. Apart from the common doctrines there were of course conflicts between various philosophical conceptions of, and corresponding programmes for, mathematics, e.g. the Leibnizian and the Kantian. And there existed almost from the beginning of mathematics the opposed philosophical doctrines of the actual and the potential infinite.

2. *Changes in the conception of geometrical intuition.* Kant defines geometry, i.e. Euclidean geometry, as a science 'which determines the properties of space synthetically and yet *a priori*' [1]. For him, as for his predecessors, there exists only one geometry and only one 'empirically real' space; and it is not their uniqueness but the synthetic *a priori* nature of the propositions of geometry which needs explanation and justification. Contrary to an often repeated opinion, the discovery that the fifth Euclidean postulate can without inconsistency be replaced by its negation, does not amount to a refutation, but to a confirmation of the thesis that it is synthetic. Nor does it amount to a refutation of the thesis that it is *a priori* in the sense of being 'independent of experience and even of all impressions of the senses' [2] since what is empirically testable is a geometry not by itself, but in conjunction with empirical hypotheses.

The discovery of non-Euclidean geometries, however, conflicts with the hitherto unquestioned assumption of the uniqueness of Euclidean geometry at least to the extent of calling it into question. The thesis of its uniqueness has therefore to be defended or abandoned. Felix Klein adopts the latter course. He notes that the perception of spatial relations is inexact and idealized in the axioms of geometrical theory: 'The results of any observations are valid only within certain limits of exactness and under particular conditions; in setting up the axioms we replace these results by statements of absolute precision and universality' [3].

---

1 *Critique of Pure Reason*, B. 40.

2 *Op. cit.*, B. 2.

3 *Math. Annalen* **50**, 1898, quoted by L. Nelson in 'Bemerkungen

Klein's thesis implies more than a contraction of the area of geometrical reality or of intersubjective intuition to the subject-matter of Euclidean geometry *minus* the fifth postulate. No more than such a contraction would be justified if the other postulates were self-evident since the independence-proof would then have merely shown that their truth cannot be extended by deduction to the fifth postulate, the truth of which, if it were true, would not be self-evident. However, Klein's thesis affects all postulates and theorems of Euclidean geometry since the distinction which he makes between the inexact observational results and the absolute precision of the axioms is justified even for such elementary notions as perceptual and mathematical betweenness and coincidence.

The thesis is not refuted by, but compatible with, the objection that even if geometries are only 'true in possible worlds', a scientist will in fact choose the geometry of that possible world which he judges to be most similar to the actual world in those respects which (for his purposes) are relevant. It is also not refuted by Nelson's objection [1] that 'every idealization presupposes an ideal' which 'cannot be borrowed from experience, since it is to be the norm for the correction of experience'. For what is in question is not that the ideal is non-empirical, but that there is only *one* ideal, namely the ideal characterized by the axioms of Euclidean geometry.

3. *Changes in the conception of set-theoretical intuition.* The philosophical controversies about the significance of the discovery of non-Euclidean geometries were at first not extended to set-theory and arithmetic. The dominant conceptions of these imply that there exists a hard core of mathematical reality or intersubjective intuition, although its philosophical analysis, and its demarcation from what are regarded as merely auxiliary and fictitious adjunctions to it differ from school to school. Moreover, the fact, acknowledged in the philosophy of geometry, that empirical individuals are not

über die Nicht-Euklidische Geometrie und den Ursprung der mathematischen Gewissheit', 1905/6, reprinted in *Beiträge zur Philosophie der Logik und Mathematik*, Frankfurt a. Main 1959, p. 34.

[1] *Loc. cit.*

sharply separated from their spatio-temporal background and that their classification yields inexact classes, is largely ignored. The discovery of the so-called limitative theorems is thus mainly regarded as lending support to a cautious attitude in the demarcation of the area of mathematical reality and as an aid in analysing it. One is compelled, however, by the recent discovery of P. J. Cohen [1] of the independence of the continuum-hypothesis, to take more seriously the apparent analogy between the subject-matters of geometry and of set-theory. In view of the large amount of philosophical literature now available on the limitative theorems, some brief remarks on the Löwenheim–Skolem theorem, Gödel's incompletability theorem and Church's theorem perhaps will suffice in this place.

According to the Löwenheim–Skolem theorem every axiomatic theory formulated within the framework of quantification-theory, if it is satisfiable at all, is satisfiable by a denumerable model. No axiomatic theory intended to express relations between non-denumerable totalities can therefore be monomorphic (categorical). The theorem thus compels a re-examination of the philosophical thesis that mathematical reality or intersubjective intuition is capable of being unambiguously and exhaustively reflected by conceptual or linguistic formulations, in particular axiomatic systems and to distinguish between reflections which are and those which are not monomorphic.

In view of this distinction we are faced with a choice between three philosophical positions: (i) the subject matter of mathematics is mind-independent or intersubjective, but only in its denumerable parts capable of monomorphic reflection by axiomatic theories; (ii) the mathematics of non-denumerable totalities is not a description of mathematical reality or intersubjective intuition but 'true in a possible world only'; (iii) not all the means for a monomorphic axiomatization of mathematics have been discovered. The philosophical relevance of the theorem consists, thus, in having at least provided this trichotomy as a new philosophical premise

---

1 See P. J. Cohen 'The Independence of the Continuum Hypothesis' in *Proc. Nat. Acad. Sci.* **50, 51,** 1963–4.

which, together with other philosophical assumptions, leads to new conclusions about the subject-matter of mathematics, e.g. a (non-Kantian, non-Platonic and non-Leibnizian) doctrine of different layers of mathematical reality.

The situation is similar as regards the philosophical significance of Gödel's incompletability theorem and Church's theorem. The former, in the strengthened form of Rosser, states that every axiomatic theory, rich enough to contain a formalization of arithmetic, is either inconsistent or contains a formula such that neither it nor its negation is provable within the theory and such that its truth can be demonstrated by extra-theoretical arguments. The latter states that quantification-theory is undecidable in the sense that the set of its theorems, though recursively enumerable, is not recursive. The theorems either compel a weakening of the conception of mathematical reality or intersubjective intuition or else a contraction of their area, if they are conceived as capable of being reflected by a complete, decidable theory constructed by such means as were available to Gödel. And the philosophically important point is just that mathematical reality was so conceived by Leibniz, by Kant, by Russell, perhaps by Plato, and with some qualifications by Hilbert. If this philosophical conception is an exaggeration, then the theorems provide new reasons for so regarding it.

Since every undecidable formula which is constructed by Gödel's methods, is true provided the system to which it belongs is consistent, it would be unreasonable to regard as equally justified the extensions of the original theory arising from adding the undecidable formula on the one hand and its negation on the other to its axioms. This is not the case with the continuum-hypothesis. The proof of its independence leads – at least from the point of view of contemporary mathematical knowledge – to a 'bifurcation' of set-theory.

In his paper 'What is Cantor's Continuum Problem?' [1] Gödel gives mathematical and epistemological reasons for rejecting the

---

[1] *Am. Math. Monthly* **54**, 1947, reprinted (with additions) in *Philosophy of Mathematics*, edited by P. Benacerraf and H. Putnam, New Jersey, 1964.

thesis that if the independence of the continuum hypothesis were to be proved 'the questions of its truth would lose its meaning, exactly as the truth of Euclid's fifth postulate became meaningless for the mathematician' [1]. As regards the mathematical objections, most of them are withdrawn in a postscript to the paper, added after Cohen's proof had become known to Gödel. I now turn to his philosophical remarks.

4. *Gödel's philosophical approach to the continuum postulate.* Gödel's philosophical remarks on the foundations of set-theory are all too rare, especially as he clearly sees the need for a 'more profound analysis (than mathematics is accustomed to giving)' of the terms and axioms of set-theory, an analysis which he regards, at least in part, as philosophical and more particularly epistemological (p. 262). His own approach is, and is acknowledged by him to be [2], similar to that of Kant. It is of some interest to compare Gödel's 'kantianism' with that of Brouwer and Hilbert. Brouwer adopts the doctrine of the Transcendental Aesthetic as regards the pure intuition of time and of mathematical existence as constructibility in pure intuition. Hilbert accepts not only the doctrine of the Transcendental Aesthetic, at least to the extent that he assumes an intuitively indubitable subject matter for synthetic (combinatorial) propositions; he accepts also the doctrine of the Transcendental Dialectic that the notions of actual, infinite totalities are Ideas of Reason. With Kant he regards these Ideas as neither abstracted from, nor applicable to, either sense-experience or intuition. And, again with Kant, he holds that propositions in which the Ideas of Reason occur as constituents can without inconsistency be adjoined to objective propositions, in which the Ideas do not occur. Since Kant regards the notion of moral freedom as an Idea of Reason, his attempt to prove the consistency of a deterministic science with the thesis that man is free, is an anticipation of Hilbert's programme of proving the consistency of mathematical theories which contain besides finitist propositions, which are infinity-free, also propositions about infinite totalities.

1 *Op. cit.*, p. 270.

2 See his footnotes 14 and 40.

What Gödel, apart from the Transcendental Aesthetic, which assumes an immediately given manifold of mathematical intuition, finds suggestive in the Kantian philosophy, is the central idea of the Transcendental Analytic, namely the constitution of physical objects from the manifold of sense-experience and intuition by the synthesis of the understanding which is the ground of the applicability of the Categories. He insists with Kant that the synthesis of a manifold, its connection into a new unity, cannot be found within the manifold itself. Where he differs from Kant is in distinguishing between, on the one hand, a specific mathematical synthesis which unifies immediately given mathematical manifolds and manifolds constituted by mathematical synthesis and, on the other, a synthesis which unifies the manifold of sense-experience into physical objects. He seems to me to be mistaken in attributing to Kant the view that the 'abstract elements in our empirical ideas', i.e. the data of mathematical intuition are 'purely subjective', especially as Kant is at great pains to insist on the empirical reality of space and time and to oppose his transcendental to any subjective idealism. It is also not quite clear whether Gödel considers every notion of set [1] or only some, e.g. the notion of higher-order-sets or of infinite sets, as unifying notions since he says (p. 271) that 'we do have something like a perception also of the objects of set-theory'.

The reason why Gödel assumes a specifically mathematical synthesis of the understanding is his assumption that 'the "given" underlying mathematics is closely related to the abstract elements in our empirical data' (*loc. cit.*) and not, as Kant held, identical with them. In this he agrees with F. Klein [2] whose implied relativism he, however, does not share. Gödel's remarks on the epistemological aspect of the continuum postulate can be fairly described as a sketch of a 'metaphysical exposition' of the concept of mathematical reality i.e. the isolation of its *a priori* features [3]. He suggests the possibility of, but does not in any way attempt, a transcendental

---

1 See Gödel's footnote 14.

2 See footnote on p. 121.

3 See, e.g., *Critique of Pure Reason*, B. 38.

deduction of the continuum postulate which would justify it as a true, synthetic proposition *a priori*.

Here, as Gödel sees quite clearly, a comparison of the philosophical implications of the independence of the fifth Euclidean postulate with those of the independence of the continuum postulate, is in place. Both are *a priori* propositions and both are by their respective independence proofs shown to be synthetic. What, however, is still in question is their uniqueness or truth, which cannot be established by reference to their applicability in any empirical theory or by psychological or pragmatic arguments. Gödel seems to suggest that we are confronted with the choice of either assuming that the continuum postulate is true or else that it is meaningless for the mathematician and thus, presumably, for everybody else. But this alternative seems to depend on an unwarranted definition of 'meaning', since propositions which are true in a possible world only, especially a world of poetic or mathematical imagination, are in an unperverted sense of 'meaning' just as meaningful as propositions which are true in *the* actual world or reality.

It would not do to argue from a supposed success of Kant's with the transcendental deduction of the Categories to the likelihood of a successful transcendental deduction of *the* unifying set-theoretical concepts since the Kantian transcendental deduction cannot be regarded as successful. This becomes clear if one attends not so much to its explicit dependence on the traditional logic as to its implicit dependence on classical physics. Thus the so-called analogies of experience which imply the indispensability to common-sense and scientific thinking of the notion of a material substance and of deterministic laws of nature are incompatible with subsequent physical theories.

One could argue, and I should be prepared to argue, that Kant has nevertheless successed in establishing some less specific principles, e.g. that any discourse about objective experience, especially any scientific discourse, requires the application of *some* Categories (concepts which unify the sensory manifold and are thus applicable to, but not abstracted from, it), even though not, on that account, any of the Kantian ones; that it requires *some* principle, or principles, of conservation, though not, therefore, the principle of the conservation of material substance; and that it requires *some* connection between various states of a system, described in a commonsense or scientific vocabulary, even though the connection might be deterministic or probabilistic. Such results of an attenuated transcendental deduction would be sufficient grounds for rejecting a positivist theory of knowledge of the Humean kind. They would not imply that commonsense and scientific thinking is constrained by *one* structure of *a priori* concepts and principles; or that the generation of objective experience from the manifold of sensory data by the synthetic function of the understanding can proceed in one way only.

But in that case Gödel's analogy between the synthesis of empirical and the synthesis of mathematical data loses its force. The fundamental set-theoretical synthesis is the operation of forming the 'set of x's', the freedom of which has, in view of the antinomies, to be constrained by so-called comprehension-axioms. The contemporary situation regarding them is explained with admirable clarity by Quine [1] who shows in particular that some of the proposed axioms though 'individually consistent' and, one may add intuitively acceptable, 'would be collectively inconsistent, and that because of this incompatibility among separately tenable cases [of the unrestricted comprehension-axiom] many radically unequivalent theories have been put forward' – there being 'no evident optimum' [2]. Quine's comparison of these theories also shows that various classifications of set-theories into finitist, intuitionist, predicative, etc. cannot be regarded as exhibiting successive consistent extensions of a common logical or intuitive core. The discovery of the independence of the continuum postulate and the axiom of choice from the other postulates of the most widely used set-theories greatly increases the multiplicity of non-equivalent set-theories. Each of them is true in a different possible world, which there is no reason to regard as the actual world or as intersubjective intuition.

Gödel in an often quoted remark (*loc. cit.*, p. 263) asserts that

---

[1] *Set Theory and its Logic*, Cambridge, Mass., 1963.

[2] *Op. cit.*, pp. 37, 38 and part III.

in so far as sets occur in the mathematics of today – i.e. as sets of integers and sets of sets . . . of integers, 'the set-theoretical concepts describe some well-determined reality, in which Cantor's conjecture must either be true or false' and that, therefore, 'its undecidability from the axioms being assumed today can only mean that these axioms do not contain a complete description of that reality'. An incomplete system of interpreted axioms cannot describe the actual or possible world completely. But even if a set of axioms is complete it does not follow therefrom that what it describes completely is the actual world. The philosopher of mathematics is, as has been pointed out already, not faced with the positivist alternative that a system of axioms either describes the actual world (reality or intersubjective intuition) or nothing at all, but in addition with a third possibility, namely that the system describes a well determined possible world. If in Gödel's phrase 'some well determined reality' the first word is emphasized, then all three possibilities are admitted, if the last word is emphasized, then only the first two. The latter reading, however, seems to express Gödel's view.

5. *On the philosophy of competitive mathematical theories.* The problem of *prima facie* conflicting mathematical theories which arose first for Euclidean and non-Euclidean geometries and later for classical and intuitionist mathematics can, as the result of recent developments, no longer be ignored even for classical set-theory and arithmetic. If numbers are construed as classes of classes, then the differences in the comprehension-axioms of any two set-theories imply differences in their class-concept and these in turn imply differences in their number-concept. The problem or uneasiness about such theories might be put roughly thus: If they describe the same subject-matter they are all incompatible with each other and if they describe a different subject-matter they are all mutually compatible. Yet neither their incompatibility on the former interpretation, nor their compatibility on the latter seems to do justice to the respects in which they agree and to those in which they compete with each other.

An obvious criterion for their agreement and competition with each other is their application or applicability to experience. This criterion would, of course, be useless if the statement that a mathematical theory is 'applicable' to experience were logically equivalent to the statement that it describes experience, since in that case all mathematical theories would again have the same subject-matter. Mathematical theories, however, do not describe experience, as was recognized by Klein, and indeed long before that by Plato, and is admitted by Gödel who holds that the mathematically given is closely related to but not identical with, the empirically given (see above). This negative characterization of the link between a mathematical theory, or at least some of its axioms and theorems, and experience has led some philosophers to the view that a mathematical theory is connected with experience only as part of the whole body of our beliefs which only as a whole is verifiable, falsifiable, confirmable or disconfirmable. Most versions of this holistic empiricism or pragmatism imply the thesis that no legitimate distinction can be drawn between mathematical and empirical propositions.

It is more in line with the use of mathematics in the transactions of ordinary life and in the sciences to see the link between mathematics and experience in the identification of some mathematical with corresponding empirical propositions within a more or less specified context or, to use Klein's words, 'within certain limits of exactness and under specified conditions'. The nature of this identification of exact mathematical and inexact empirical propositions is perhaps clearest in the case of the application of geometry. It is here in particular obvious that the identificatory statements which relate a geometrical to an empirical proposition are themselves empirical [1].

If g_1 and g_2 are two geometrical propositions which cannot both be true in the same possible world, then we may call them 'coidentifiable' within certain contexts (e.g., for all contexts so far examined) if, and only if, they are within these contexts identifiable with

[1] For a more detailed discussion of the grounds of mathematical exactness and empirical inexactness in the logical frameworks underlying mathematical and empirical discourse and for a closer analysis of the structure of identificatory statements see 'An Empiricist Justification of Mathematics' in *Proceedings of the 1964 International Congress of Logic, Methodology, and Philosophy of Science.*

the same empirical, or observational, propositions. They conflict with each other with respect to a context of identification in so far as they are not coidentifiable in it. These notions of coidentifiability and conflict can be easily extended from pairs of geometrical propositions to geometries. Coidentifiability is clearly quite different from isomorphism. Two isomorphic theories – e.g., two such theories with an altogether different subject-matter – need not be coidentifiable; and two coidentifiable theories – e.g., the geometry of a Euclidean circle and of a regular polygon with a very large number of sides – need not be isomorphic.

The reasons for this account of the agreement and conflict between geometries which are true in different possible worlds were twofold, namely the multiplicity of the theories and the contrast between the idealizing geometrical propositions and their idealized empirical counterparts. About the multiplicity of set-theories and arithmetics there can, as we have seen, be no doubt. The one generally accepted arithmetic, which is intuitively clear even to young children, is nothing but a set of empirical propositions with respect to which parts of all the different arithmetical theories, intuitionist or classical, are within the contexts of practical life, coidentifiable. If these different arithmetics had one and the same (mathematical or empirical) subject-matter they would be mutually incompatible so that at most one of them would be true.

A little reflection will show that the contrast between idealizing mathematical and idealized empirical propositions holds also for the most elementary set-theoretical and arithmetical notions. The sets of set-theory are exactly demarcated and not inexact or 'open-textured' as are classes of empirical individuals. Again empirical individuals have a spatio-temporal background from which they are not sharply separated so that the answer when and where such an individual begins or ends admits of no precise answer. Mathematical individuals, e.g. individual numbers, have no spatio-temporal background and are precisely distinguishable from each other. The principle of mathematical induction is backed by the availability without limit of new mathematical, but not also of empirical, individuals, etc.

Disregarding always possible appeals to some dogmatic metaphysics, one must conclude that the metamathematical discoveries of the present century imply the falsehood of the common doctrines shared by the classical philosophies of non-competitive mathematical theories, but not that they imply the positivist doctrine that if mathematics is not true-or-false in the actual or every possible world, it is therefore meaningless. They support on the other hand the view that the various unequivalent, mutually compatible theories are meaningful, i.e. true in different possible worlds, even though none of them is actual, and that their agreement or conflict with each other lies in their relation to experience, namely the coidentifiability or otherwise, of at least some of their axioms or theorems with empirical propositions.

It is, perhaps, of some interest to compare the above account with others which seem to be in vogue. The positivist doctrine of the 'meaninglessness of mathematics' has already been discussed. Most of the other accounts are based on a conflation of 'applicability' in the strong sense of description and in the weak sense of description or identification of idealizing with idealized propositions (of the application in the strong sense of concepts to their substitution instances and their application in the weak sense to entities which may, but need not, be their substitution instances). Platonism recognizes that mathematical propositions are idealizations of empirical propositions, but ignores the multiplicity of different coidentifiable idealizations and asserts that one of them is true of the mind-independent reality. Conceptualism differs from Platonism in substituting for the one mind-independent reality one mind-dependent, intersubjective intuition. Realism, materialism, and nominalism regard all or some mathematical propositions as descriptive of experience because they do not, as regards these propositions, distinguish between applicability in the strong and in the weak sense (between the application of mathematical concepts to their substitution-instances and their application *via* an idealization to empirical objects). The decision between the account here proposed and any of the others is unlikely to affect the development of mathematical thinking. But the task in hand was to discuss philosophical implications of mathematics not mathematical implications of philosophy.

DISCUSSION

Gert H. Müller: *An old philosophical question – and the recent results in the foundations of mathematics.*

The history of philosophy and an analysis of the structure of sciences show that there is a remarkable difference *and* a deep connection between natural and formal sciences (Mathematics and Logic). In considering formal sciences we find an analogous difference *and* connection between Mathematics and Logic. There have been attempts to describe or to characterize this antagonistic relation in a dualistic or monistic way. In the latter case either the one or the other part dominates (empiricism versus rationalism) or some foreign principles are used (Kant). In practically all these attempts since Greek times the assumption was made that there is only *one* formal science or respectively only *one* Logic. These assumptions have become doubtful in view of the results of foundational research in Mathematics and Logic in the last 150 years. Körner is therefore right in speaking of a fundamental change in the philosophical question concerning their antagonistic relation. The unity of formal science or Logic is a problem today, and not a self-evident fact.

From the Greek way of thinking we have taken over an ideal of the sciences which does not accept any plurality as definitive. This ideal has the character of a 'regulative idea' implying the unity of all sciences, and it does not seem likely that we shall abandon this ideal. Now I think that one can use Körner's concept of coidentifiability as a starting point for a philosophical treatment of the above-mentioned antagonistic relation, even if one rejects or weakens the classical assumption of the unity of formal science.

Certainly there are considerable difficulties in making Körner's concept of coidentifiability precise, but it is clear that well-known examples in the sciences can help us to understand this concept. Consider, for example, the compatibility of Euclidean and non-Euclidean geometry when they are applied in small (e.g. terrestrial) domains, of Newtonian and Einsteinian mechanics in small domains (i.e. planetary systems, short times), of classical and quantum mechanics applied to 'big' masses, and so on. We further recognize such a compatibility between classical and intuitionistic logic relative to finite domains, of classical and intuitionistic number theory relative at least to primitive recursive functions, and so on.

In considering these statements of compatibility we see that they are all relative to a *domain of knowledge* which is external to the theories compared. (Körner also stresses this connection with the given context in his introduction of the concept of coidentifiability, cf. p. 130). Such domains of knowledge are in general determined by the source of knowledge, or perhaps better by a domain of familiarity (*Vertrautheitsbereich*), such as our usual sense-experience, by the more or less imaginable finiteness, by mathematical induction, and so on. These domains of familiarity are then used in judging theories; first of all in order to refute some theories, but also in order to define some domains of coidentifiability.

(It may be remarked that we do not prejudge in advance the strong distinction between formal and empirical science, i.e. between the analytic and the synthetic. This does not mean that the domains of familiarity are considered as absolute or unanalysable.)

What is called 'the ontological' is indeed in the first place a 'regulative Idea'. Acquaintance with some ontological domains we get only by means of domains of familiarity and their role in judging alternatives. An ontological conflict between two theories (for instance Euclidean and non-Euclidean geometry or classical and intuitionistic number theory) is then excluded, since it is excluded in the domain of coidentifiability determined by the given domain of familiarity; or in other words, there is by definition no ontological commitment of theories outside the domains of familiarity. Clearly there are extensions of the ontological commitment of theories (for instance in geometry) in view of the growth of our knowledge (for instance in cosmology), but this means that our domains of familiarity are extended, too.

In mathematics we define 'possible' structures with the help of the usual postulational method. So long as we are interested in algebraic theories a certain richness of models is desirable, and the domain of models is defined by some set theory. But if we are interested in the theory of numbers, the continuum, or the theory of sets, the situation changes, since we believe we have some previous knowledge by intuition (imagination) of number, the continuum, or sets, and we realize that in these cases a certain standard form of applicability is a *defining* property of numbers, of the continuum, of sets. Here again we use some domains of familiarity (some sources of knowledge) to judge different postulated theories about numbers and so on. For instance 'previous knowledge' or familiarity with 'the inductive' can be used to delimit classical number theory and possibly a denumerable analysis (in the sense of Lorenzen, Schütte, Wang). Modifying the methods by which theorems of number theory are proved one gets the known difference between intuitionistic and classical number theory in a natural way. Clearly one can construct a deep ontological conflict between these two theories of numbers, but only if one uses the concept 'ontological' in a sense which is much stronger than the principles which are used to compare these two theories.

On the basis of geometrical intuition (as P. Bernays sometimes maintains) one can define a domain of familiarity which assures us of the existence of the continuum; and one gets the 'same' continuum using combinatorial methods (Cantor). But as regards an arithmetical approach to the continuum (*via* Cantor's second number class) we only know that it *can* be identified with the combinatorial one (Gödel) but that it need not be so identified (Cohen). True, in higher set theory there seems to be a deep difference between the combinatorial approach (*via* postulates about Boolean Algebras) and the arithmetical, iterative one. However, our domains of familiarity in contemporary Mathematics are too small to pass judgment on the comparative merits of different postulates in these cases.

Clearly some 'global' hypothesis may also have consequences in the domains of familiarity. But so far it has always been possible to overcome such difficulties either by appealing to the priority of a domain of familiarity or by becoming familiar with something new.

Y. Bar-Hillel: *On a neglected ontology-free philosophy of mathematics.*

It is interesting that Körner, who has been so meticulous in mentioning in his talk a good number of recent philosophies of mathematics, viz. platonism, conceptualism, realism, materialism, nominalism and positivism, in addition to his own view, should have neglected to mention what in my opinion is the most adequate philosophy of mathematics, viz. the strictly non-ontological conception of mathematics by Carnap, Kemeny and others. (As a matter of fact, Carnap's name is not even mentioned in Körner's discussion at all.) Their philosophy of mathematics is certainly not a branch of what Körner calls 'positivism' since they would not want to call mathematics 'meaningless', though they would indeed insist that arithmetic, analysis, set-theory, etc. (but not geometry) are disciplines that are in no sense directly interpretable in observational terms. Their 'relation to experience' is neither direct interpretability, nor idealization, nor 'coidentifiability' but, exactly like that of logic, indirect interpretation which for most formalized arithmetical and set-theoretical systems has turned out to be incomplete. I admit that there exists at present no satisfactory detailed statement of their ontology-free conception and can only hope that someone will sooner or later come up with one.

I would therefore want to insist that, in addition to the three philosophical positions mentioned by Körner as to the subject-matter of mathematics, another one deserves attention, namely the one according to which mathematics has *no* subject-matter (which, of course, does not mean that it is 'meaningless', any more than the fact that logic has no subject-matter – and I hope that Körner would agree to this formulation – means that logic is meaningless). In addition to a many-layered doctrine of mathematical reality, let me recall to your attention that a zero-layered doctrine is in competition.

S. Körner: *Reply.*

I am grateful to Professors Müller and Bar-Hillel for their helpful comments. Professor Bar-Hillel objects to my not having discussed

Carnap's 'ontology-free' philosophy of mathematics. An explanation, if perhaps not an excuse, is that I have only fairly recently criticized some of Carnap's general doctrines, which are closely related to his philosophy of mathematics. These are his distinction between cognitively meaningful and meaningless statements, his analytic-synthetic dichotomy, his conception of explication and his account of the rules of correspondence between theoretical and observational language [1]. However, according to my account of them, mathematical theories are also ontologically uncommitted, namely in the sense that mathematical concepts are not instantiated in sense-experience, in the physical world or in some other reality. To apply mathematical concepts to experience is to identify them with empirical concepts which are instantiated in it. As idealizations, which are 'true in possible worlds only', mathematical theories carry no heavier ontological burdens than do works of literary imagination.

I have much sympathy with Müller's suggestions and hope that he will find an occasion for developing them independently at length. There are important analogies within mathematical (and other) theories to the logical disconnection between observational and mathematical concepts on the one hand and to their limited identifiability on the other [2]. Müller's notion of an – empirical or non-empirical – 'domain of familiarity' (*Vertrautheitsbereich*), with respect to which various idealizations are within limits coidentifiable, seems to me useful not only in exploring these analogies, but also the manner in which the scope of mathematical (and other) theories is extended. It may, however, be desirable and possible to define the notion in a less 'psychologistic' manner.

---

1 See my review article on Schilpp (ed.), *The Philosophy of Rudolf Carnap*, in *Mind*, 1966.

2 See e.g. ch. XI of *Experience and Theory*, London 1966.

INFORMAL RIGOUR AND COMPLETENESS PROOFS

GEORG KREISEL
Stanford University

It is a commonplace that formal rigour consists in setting out formal rules and checking that a given derivation follows these rules; one of the more important achievements of mathematical logic is Turing's analysis of what a formal rule is. Formal rigour does *not* apply to the discovery or choice of formal rules nor of notions; neither of basic notions such as *set* in so-called classical mathematics, nor of technical notions such as *group* or *tensor product* (technical, because formulated in terms of an already existing basic framework).

The 'old fashioned' idea is that one obtains rules and definitions by analyzing intuitive notions and putting down their properties. This is certainly what mathematicians thought they were doing when defining length or area or, for that matter, logicians when finding rules of inference or axioms (properties) of mathematical structures such as the continuum. The general idea applies equally to the so-called realist conception of mathematics which supposes that these intuitive notions are related to the external world more or less as the number 4 enters into configurations consisting of 4 elements, and to the idealist conception which denies this or, at least, considers this relation as inessential to mathematics. What the 'old fashioned' idea assumes is quite simply that the intuitive notions are *significant*, be it in the external world or in thought (and a *precise* formulation of what is significant in a subject is the result, not a starting point of research into that subject).

Informal rigour wants (i) to make this analysis as precise as possible (with the means available), in particular to eliminate doubtful properties of the intuitive notions when drawing conclusions about them; and (ii) to extend this analysis, in particular not to leave undecided questions which can be decided by full use of evident properties of these intuitive notions. Below the principal emphasis is on intuitive notions which do not occur in ordinary mathematical practice (so-called new *primitive notions*), but lead to new axioms for current notions. We give three (or, if one includes App. B which was added after the talk, four) applications, mostly following the 'old fashioned' idea of pushing a bit farther than before the analysis of the intuitive notions considered. Section 1 concerns the difference between familiar independence results, e.g. of the axiom of parallels from the other axioms of geometry, on the one hand and the independence of the continuum hypothesis on the other; the difference is formulated in terms of higher order consequence. Section 2 deals with the relation between intuitive logical consequence on the one hand and so-called semantic resp. syntactic consequence on the other. Section 3 analyzes Brouwer's 'empirical' propositions in his proof of

$$\sim \forall \alpha [\sim \sim \exists x(\alpha x = 0) \supset \exists x(\alpha x = 0)].$$

Sections 2 and 3 affect completeness questions for classical and intuitionist predicate logic, which accounts for the title of this talk; quite generally, problems of completeness (of rules) involve informal rigour, at least when one is trying to decide completeness with respect to an intuitive notion of consequence.

Appendix A contains an axiomatic theory of definable properties which is used in section 2; it is of independent interest because it formulates a general relation, which includes as a particular case the well known passage from sets to classes.

Appendix B was stimulated by A. Robinson's lecture at this meeting on non-standard models. There are two more or less independent parts. First, one asks whether standard or non-standard models are more fundamental. We propose a criterion in terms of *definability* (as is to be expected in this old fashioned lecture: in favour of standard models). Secondly we consider a question, proposed by A. Robinson, because the question is a beautiful example of an *informal derivation of axioms*: Granted the existence of non-standard models, certain axioms extending the axioms of current first order arithmetic, are evident; we show that they are a conservative extension.

(0) *The case against informal rigour* (or: antiphilosophic doctrines). The present conference showed beyond a shadow of doubt that several recent results in logic, particularly the independence results for set theory, have left logicians bewildered about what to do next: in other words, these results do not 'speak for themselves' (to these logicians). I believe the reasons underlying their reaction, necessarily also make them suspicious of informal rigour. I shall try to analyze these reasons here.

(a) *Doctrinaire objections* (*pragmatism, positivism*). Two familiar objections to informal rigour are these:

(i) Why should one pay so much attention to intuitive notions? What we want are definitions and rules that are *fruitful*; they don't have to be *faithful* to notions that we have already. One might perhaps add: these notions are formed without highly developed experience; so why should they be expected to be fruitful?

Besides this (pragmatic) objection we have a more theoretical (positivist) objection.

(ii) These intuitive notions, in particular the (abstract) notions of validity, set, natural number or, so as not to leave out intuitionism, intuitively convincing proof, are illusions. When one examines them one finds that their solid content lies in what we do, in how we act; and, in mathematics, this is contained in the formal operations we perform.

A certain superficial plausibility cannot be denied to these objections. First, when some abstract intuitive notion turns out to be equivalent, at least in a certain context, to a positivistic relation, i.e. one definable in particularly restricted terms, this has always important consequences. For instance (for detail, see section 2) logical consequence applied to first order formulae, is equivalent to formal derivability; and first order axiom systems permit a more general theory than higher order systems. Consequently, at a particular stage, the (pragmatically) most rewarding work in the subject may consist quite simply in exploiting the discovery of such an equivalence. Second, one may be impressed by the slow progress of work on some of the intuitive notions, particularly those associated with traditional philosophic questions: pragmatism discourages such work, and positivism tries to give theoretical reasons for the slow progress. Now, objectively, such a negative attitude is not supported by the facts because progress was also slow in cases where decisions *were* eventually obtained. (About 30 years between Hilbert's first formulation of his finitist programme, cf. (c), page 146, and Gödel's incompleteness theorems; nearly a further 30 years till a precise analysis of finitist proof was attempted.) But, subjectively, if a particular person *is* discouraged by the slow progress he had surely better find himself another occupation. Certainly, scientifically speaking, one is in a wholly futile position if one finds oneself stuck both with philosophy as a profession and with antiphilosophical views such as pragmatism or positivism (perhaps, after having been attracted by traditional questions in one's youth). For, having repudiated specifically philosophical notions one is left with those that are also familiar to specialists in other fields: what jobs can one hope to do as well as these specialists? including the jobs of clarification or explication (if they are to be done in current terms)? I think this futility is felt quite consciously by many of the people involved.

Having granted all this: what is wrong with (i) and (ii)? Quite simply this. *Though they raise perfectly legitimate doubts or possibilities, they just do not respect the facts, at least the facts of actual intellectual experience.* This is particularly irritating because pragmatism pays so much lip service to experience, and positivism claims to be empirically minded.

Ad (i). Let us even take for granted that we know roughly what is fruitful; after all, here again (i.e. as in the case of significance) a precise formulation may only be possible after a good deal of experience. Perhaps we do not know a general reason why intuitive notions *should* have stood the tests of experience well; as one sometimes says: they might not have done so. But the fact remains that they, or, at least, many of them, have. Reflection shows that we certainly couldn't have what we understand by 'science' if they hadn't. Instead of trying to find reasons for, or limitations of, this superficially remarkable situation, (i) denies its existence! (a most unpragmatic pragmatism). Two related so-called pragmatic principles are to be mentioned. One says that one must treat each problem 'on its merits'; one might have to; but taken literally this would leave little room for general theory or for the distinction between what is fundamental and what is secondary. Or again (in mathematics), one sometimes criticizes complacently 'old fashioned' disputes on the right definition of measure or the right topology, because there are several definitions. The most striking fact here is how *few* seem to be useful: these haven't dropped from heaven; they, obviously, were formulated *before* their applications were made, and they were not generally obtained by trial and error. If they had been so obtained, mathematicians shouldn't be as contemptuous as they are about the study of little variants in definitions. Similar remarks apply to the choice of axioms; but since this is of direct logical importance the subject will be taken up in its proper place in (b).

Ad (ii). Clearly, if, consciously or unconsciously, one insists on analyzing the 'solid content' in positivistic, in particular, formal, terms this is what one will find. Though more specific points about formalism and formalisation are taken up throughout this paper in particular (c) and section 2 below, some matters of principle are in order here. It might have turned out that the notions which present the most serious difficulties in practice are indeed abstract ones. But, quite naively, this is not so: knowing whether two inscriptions *mean* the same is often no harder than knowing if they *read* the same! Equally, as was mentioned on p. 140 above, sometimes it does turn out that some notions are fully represented in formalistic terms: but this has to be verified and section 1 shows limitations. Perhaps one should distinguish between formalism (and positivism), which is merely a negative antiphilosophical doctrine, and a *mechanistic* conception of reasoning ('mechanism' in the sense of Turing), which would lead one to expect a full formalistic analysis of actual reasoning. It is to be remarked that, so far, the most that has ever been shown in support of this conception is that in certain areas (e.g. elementary logic, section 2) reasoning *could* be mechanical in the sense that a mechanism would get the same results; not that it *is*, i.e. that it would follow the same routes. It may be that the mechanistic conception is the only moderately clear idea of reasoning that we have at present.

But a good positivist should not conclude from this that therefore this idea is correct.

(b) *Unreliability of some intuitive notions; the role of formalisation in their analysis.* A much more serious point than the portemanteau objections (i) and (ii) under (a) concerns specific abstract notions, for instance – to take the most famous example – the notion of set. *Have not the paradoxes shown the complete unreliability of our intuitive convictions at least about this particular notion?*

First of all, historically speaking, this couldn't be farther from the truth! Wasn't Cantor a misunderstood martyr in the face of widespread reactionary prejudice against employing the notion of set (or, as it was then called: class) in mathematics? If so, the paradoxes supported the intuitive convictions of those reactionaries.

It is probably true to say that the reactionary caution was due to this: *class* presented itself as a vague notion, or, specifically, a mixture of notions including (i) finite sets of individuals (i.e. objects without members), or (ii) sets *of* something (as in mathematics, sets of numbers, sets of points), but also (iii) properties or *intensions* where one has no *a priori* bound on the extension (which are very common in ordinary thought but not in mathematics). If we are thinking of sets *of* something, e.g. of objects belonging to a, then the comprehension axiom is to be restricted to read (for any property P)

$$\exists x \forall y (y \in x \leftrightarrow [y \in a \ \& \ P(y)]);$$

but if we are thinking of properties, given in intension, whose range of definition is not determined, we may well have (with variables ranging over properties)

$$\exists Q \forall R [Q(R) \leftrightarrow P(R)]:$$

only one had better remember that these properties are not everywhere defined and so the laws of two valued logic are not valid. (So, to be precise, the logical symbols have different meaning in the two cases.)

Now, the reactionaries were wrong because at least one element of the mixture (namely: set *of* something), first described clearly by Russell and, especially, Zermelo, has proved to be marvellously clear and comprehensive. But before this analysis the prospects were not rosy: what one could hope to do was to put down *assertions which are satisfied by all elements of the mixture*, and it just didn't look as if this 'common part' was going to lead to a mathematically rich theory; more precisely, to something in terms of which anything like (then) current mathematics could be interpreted. In contrast, axioms which are evidently valid for the particular notion isolated by Zermelo (cumulative type structure) give a formal foundation (even) for a great deal of present day mathematical practice.

The main problem which, in my opinion, the paradoxes present can be put quite well in old fashioned language: *what are the proper laws* (the 'logic') *satisfied by the intensional element of the crude mixture*, in particular, the element (iii) which satisfies the unrestricted comprehension axiom? [1]

Two *conclusions* about matters discussed at this meeting follow. First, Zermelo's analysis furnishes an instance of a rigorous *discovery of axioms* (for the notion of set). To avoid trivial misunderstanding note this: What one means here is that the intuitive notion of the cumulative type structure provides a coherent *source* of axioms; our understanding is sufficient to avoid an endless string of ambiguities to be resolved by further basic distinctions, like the distinction above between abstract properties and sets *of* something. Pragmatically speaking, cf. (a) above, one does not have to put up with an *ad hoc* collection of different axioms for different 'purposes' (though *ad hoc* considerations may be needed to show which of the new axioms are relevant to a particular purpose). Denying the (alleged) *bifurcation* or *multifurcation of our notion of set of the cumulative hierarchy* is nothing else but asserting the properties of our intuitive conception of the cumulative type-

---

[1] The recursion theorem for partial (recursive) functions is analogous to such a comprehension axiom: this might serve as a model 'in the small' for *very* abstract notions which again satisfy a comprehension axiom without type restriction. It should be mentioned that the views of the preceding two paragraphs contradict those (implicit) in the literature; e.g. Mostowski, in these Proceedings, sees the kernel of set theory in a kind of common part of different notions of set, and Rasiowa–Sikorski [6] regard the paradoxes as a dead (fruitless) issue.

structure mentioned above. This does not deny the established fact that, in addition to this basic structure, there are also technically interesting non-standard models; cf. App. B, defined in terms of the basic structure. – From the present point of view, the importance of adding strong *axioms of infinity* (existence of large ordinals α) is clear. For, adding them does not restrict the unconditional assertions we can make about initial segments of the hierarchy below α; but leaving them out stops us from saying anything unconditional beyond α. For instance, consider the axiom of infinity (existence of ω, which is included in Zermelo's axioms); an arithmetic theorem A proved in Zermelo's theory, an arithmetic theorem B proved from the Riemann hypothesis only. Without the assumption of ω, both A and B are hypothetical: 'statistically', I suppose, 'equally' justified, since, after all, the Riemann hypothesis has not led to a contradiction (otherwise it would be refuted). But, intuitively, A is established and B is not. Unless one denies the validity of this distinction, leaving out the assumption of ω conflicts with requirement (ii) on informal rigour on pp. 138 ff.

Second, the actual *formulation* of axioms played an auxiliary rather than basic role in Zermelo's work: the intuitive analysis of the crude mixture of notions, namely the description of the type structure, led to the good axioms: these constitute a record, not the instruments of clarification. And a similar conceptual analysis will be needed for solving the problem of the paradoxes.

(c) *Formalisation.* What has been said above about the formulation of axioms, applies even more to the formulation of rules of inference (for further details, see section 2). In fact, on quite *general* grounds, one would expect the role of formalisation to be always auxiliary in the analysis of notions. After all, the job of formalisation is to record and codify arguments without distinguishing the good from the bad. And, leaving generalities, if one considers the so-called 'crises' in mathematics, one never disagreed about the inferences themselves, but either about the axioms (comprehension axiom) or about the *rules* of inference (law of the excluded middle in the intuitionistic criticism). So, *precision of the notion of consequence was not a primary issue.* If one believes that precision is the principal objective of formalisation, and formalisation of logic, one cannot be surprised that French mathematicians used to think of logic as the hygiene of mathematics and English mathematicians of logic as a matter of dotting the *i*'s and crossing the *t*'s. (What logic does is to study notions which were previously not recognized at all, or, if recognized, only used heuristically, and not made an object of detailed study; among them traditional philosophic notions.)

It seems certain that, for the psychological process of understanding, formalisation is indeed important; only I do not have a good analysis. But there are some quite clear reasons why the role of formalisation for *fundamental* analysis should be overvalued.

First and foremost, there are probably quite a few who scorn Hilbert's programme, but hold on to formalisation as a kind of collective reflex. Hilbert wanted to *show* (as the positivist in (a) above should have done) that one really lost nothing by confining oneself to formal operations, and he found a way of expressing this in the form of a mathematical problem, namely his programme. Axiomatisation, in fact strict formalisation of inference, was *essential* for the very formulation of this programme. Incidentally, there was no idea of *rejecting* on general grounds the notion of second order consequence (section 2), but of showing its equivalence, at least in suitable contexts, to formal derivability, as had been done for first order logical consequence. Unless one had been convinced of the latter equivalence one would never have engaged in Hilbert's programme.

Second, explicit formulation of axioms and rules undoubtedly plays a big role in everyday work of logicians. *Examples*: (i) If the basic concepts are accepted, one can make deductions *from* axioms clearer by eliminating those formally unnecessary to the conclusion; but, e.g. in Zermelo's analysis, to make the meaning of the basic concepts clear, he made them *more* specific. (ii) Suppose one wants to explain why a certain question happens to be open; one guesses a formal system, i.e. properties of the intuitive notion, which mathematicians are likely to use; one supports this guess by showing that current mathematics follows from these axioms, and one explains the situation by showing that this question is not decided in the formal system (I personally like this sort of thing). (iii) If one wants to remember a proof one remembers the axioms used in the proof: the less there are the less proofs have to be remembered.

Of course all this goes to show that a lot of the everyday work of logicians is not concerned with fundamental analysis at all (even if one would like it to be). As someone (nearly) said at Balaclava: *C'est magnifique, mais ce ne sont pas les fondements.*

1. *Higher order axiomatisations and independence proofs.* This section takes the precise notion of set (in the sense of the cumulative [1] type structure of Zermelo, (b), page 144) as starting point and *uses* it to formulate and refine some intuitive distinctions. Specifically, one analyses concepts according to the *order* of the language needed to define them. The connection between this technical exposition (in the sense of p. 138) and informal rigour appears in (c) below.

It is clear that the matter is technical because the very idea of a definition requires an interpretation of the language, for instance of the logical symbols. In the present case they are interpreted by means of the (set-theoretic) operations of complementation, union and projection. The languages used are those of *predicate logic* (for instance, in Church's book.)

More or less familiar examples. Such notions as *equivalence relation* or *order* are defined by first order formulae A_E, A_O resp., i.e. in ordinary predicate calculus, in the sense that a structure consisting of a domain a and a relation b on a, (i.e. $b \subset a \times a$)

---

[1] Professor Bar-Hillel asked in discussion if one starts with a structure containing *Urelemente*, i.e. objects which have no elements but are different from $\emptyset$. Evidently, the answer is yes, because most concepts do not present themselves as concepts of sets at all (apples, pears). But it is a *significant theorem* that the classical structures of mathematics occur already, up to isomorphism, in the cumulative hierarchy *without* individuals. For the reduction of mathematics to set theory it is important to convince oneself that intuitively significant features are invariant under isomorphism, or, at least, classes of isomorphisms definable in set-theoretic terms, e.g. recursive ones.

satisfies A_E, A_O if and only if b is an equivalence relation, resp. an order on a. We are nowadays very familiar with this but it was a not at all obvious discovery that intuitive notions of equivalence or order could be defined in terms of such a simple language as predicate logic of first order. Equally it was a discovery that the only concepts definable in *this* way which are unique up to isomorphism are finite structures. So, both the expressive power and the limitations of first order language came as a surprise.

The familiar classical structures (natural numbers with the successor relation, the continuum with a denumerable dense base etc.) are definable by *second* order axioms, as shown by Dedekind. Zermelo showed that his cumulative hierarchy up to ω or $\omega+\omega$, or $\omega+n$ (for fixed n) and other important ordinals is equally definable by second order formulae. Whenever we have such a second order definition there is associated a *schema* in first order form (in the language considered): For instance, in Peano's axiom

$$\forall P[\{P(0) \,\&\, \forall x[P(x) \rightarrow P(x+1)]\} \rightarrow \forall x P(x)]$$

one replaces the second order quantifier P by a list of those P which are explicitly defined in ordinary first order form (from $+$ and $\times$, for instance). A moment's reflection shows that the evidence of the first order axiom schema derives from the second order schema: the difference is that when one puts down the first order schema one is supposed to have convinced oneself that the specific formulae used (in particular, the logical operations) are well defined in any structure that one considers; this will be taken up in (b) below. (*Warning.* The choice of first order schema is not uniquely determined by the second order axioms! Thus Peano's own axioms mention explicitly only the constant 0 and the successor function S, not addition nor multiplication. The first order schema built up from 0 and S is a weak, incidentally decidable, subsystem of classical first order arithmetic above, and quite inadequate for formulating current informal arithmetic. Informal rigour requires a much more detailed justification for the choice of $+$ and $\times$ than is usually supplied.)

An interesting example of a concept that needs a third order definition is that of measurable cardinal [2]. Such concepts are rare;

for, whenever theoretically a whole hierarchy presents itself, in practice one only uses the first few levels – or a notion outside that hierarchy altogether; two concepts may be mentioned here which are not definable by any formula in the whole hierarchy of languages of predicate logic.

Evidently, neither the cumulative type structure itself nor the structure consisting of the ordinals with the ordering relation is definable by any formula of finite (or transfinite) order: for (at least, usually) one requires that the universe of the structures considered be a set, and no set is isomorphic to the totality of all ordinals, let alone of all sets; cf. also App. *A*.

NB. The (somewhat crude) classification in terms of *order* of the language considered has recently been refined by the use of *infinite* formulae, for instance, infinite first order formulae. These or, at least, important classes $\mathscr{C}$ of them are *intermediate* between first order and second order formulae since any structure definable by a formula in $\mathscr{C}$ is also definable by means of a (finite) second order formula, but not conversely. The ordinal ω is an example.

It is clear that what is achieved in the case of ω is a technical analysis of certain sets of integers by means of the notion of integer (which is used essentially in the theory of infinite formulae) – perhaps, as one uses induction to define, and obtain results about, prime numbers. The notion of integer itself is not analysed in this way.

What is much more interesting than this obvious remark is the fact that a rich theory of infinite formulae can be developed: specifically, many useful theorems about finite first order formulae can be extended to infinite ones (cunningly chosen) but not to finite or infinite second order formulae.

(a) A *reduction* of assertions about higher order consequence to first order statements in the language of set theory. Since the notion of realisation of a formula (or, of model) of any given order is formulated in terms of the basic notion of set one may expect that e.g. $A \vdash_2 B$, i.e. B is a consequence of the second order formula A, is expressed by a first order formula of set theory. More precisely, expressed by such a formula when the quantifiers are interpreted to range over all sets of the cumulative type structure. One expects this, simply because it is always claimed that this first order language is adequate for all mathematics; so if it weren't adequate for expressing second order consequence, somebody would have noticed it [1].

A simple calculation verifies this; moreover, *the definition uses exactly the same basic notions as that of first order consequence*: only instead of a quantifier $\forall a$ (over all sets) followed by a formula whose quantifiers are restricted to a, i.e. $(\forall b \in a)$, one has a formula containing also quantifiers $(\forall b \subset a)$.

As a corollary, any conclusion that we may formulate in terms of second order consequence can also be formulated by means of a first order assertion about the cumulative type structure. However, *heuristically*, that is for finding the first order assertion, it may be very useful to think in terms of second order assertions.

Example. Let $\mathscr{Z}$ be Zermelo's axiom with the axiom of infinity, and let CH be the (canonical) formulation of the continuum hypothesis in the following form: if C_ω is the collection of hereditarily finite sets without individuals, $C_{\omega+1} = C_\omega \cup \mathfrak{P}(C_\omega)$, $C_{\omega+2} = C_{\omega+1} \cup \cup \mathfrak{P}(C_{\omega+1})$, CH states that

$$X \subset C_{\omega+1} \rightarrow (\overline{\overline{X}} \leqslant \overline{\overline{C}}_\omega \vee \overline{\overline{X}} = \overline{\overline{C}}_{\omega+1}),$$

which is expressed by means of quantifiers over $C_{\omega+2}$. As Zermelo pointed out (see above), if we use the current set-theoretic definition $Z(x)$ of the cumulative hierarchy, in any model of $\mathscr{Z}$, this formula Z defines a C_σ for a limit ordinal $\sigma > \omega$. Consequently we have

$$(\mathscr{Z} \vdash_2 CH) \vee (\mathscr{Z} \vdash_2 \sim CH).$$

Note that CH is formulated in first order language of set theory.

(b) *Distinctions* formulated in terms of higher order consequence. In contrast to the example on CH above, Fraenkel's

---

[1] One cannot be 100 per cent sure: for instance, consider the so-called truth definition. We have here a set T of natural numbers, namely Gödel numbers $\bar{\alpha}_i$ of first order formulae of set theory, such that $n \in T \leftrightarrow \exists i (n = \bar{\alpha}_i \;\&\; \alpha_i)$, i.e. T is defined by

$$(n = \bar{\alpha}_1 \;\&\; \alpha_1) \vee (n = \bar{\alpha}_2 \;\&\; \alpha_2) \vee \ldots$$

As Tarski emphasised, T is not definable by means of a first order formula (in the precise sense above).

replacement axiom is not decided by Zermelo's axioms (because $\mathscr{Z}$ is satisfied by $C_{\omega+\omega}$, and Fraenkel's axiom not); in particular it is independent of Zermelo's second order axioms while, by Cohen's proof, *CH* is only independent of the *first order schema* (associated with the axioms) of Zermelo–Fraenkel.

This shows, first of all, the (mathematical) fact that *the distinction between second order consequence and first order consequence* (from the schema) *is not trivial.*

Secondly, it shows a *difference* between the independence of the axiom of parallels in geometry on the one hand and of *CH* in first order set theory. In geometry (as formulated by Pasch or Hilbert) we also have a *second order* axiom, namely the axiom of continuity or Dedekind's section: *the parallel axiom is not even a second order consequence of this axiom,* i.e. it corresponds to Fraenkel's axiom, not to *CH*.

Finally, consider the empirical fact that nobody was astonished by the independence of Fraenkel's axiom, but many people were surprised by Cohen's result. This reaction is quite consistent with my assertion above that the evidence of the first order schema derives from the second order axiom. Even if one explained to a mathematician the distinction above he would marvel at the ingenuity required to *exploit* it; for, in his own work he never gives a second thought to the form of the predicate used in the comprehension axiom! (This is the reason why, e.g., Bourbaki is extremely careful to isolate the assumptions of a mathematical theorem, but never the axioms of set theory implicit in a particular deduction, e.g. what instances of the comprehension axiom are used. This practice is quite consistent with the assumption that what one has in mind when following Bourbaki's proofs is the second order axiom, and the practice would be horribly unscientific if one really took the restricted schema as basic.)

(c) Connection between *informal rigour and the notion of higher order consequence.* The first point to notice is that this notion is needed for the very formulation of the distinction above. This illustrates the weakness of the positivist doctrine (ii) in (a), page 140, which refuses to accept a distinction unless it is formulated in certain restricted terms. [NB. Of course *if* one wants to study the formalist reduction, Hilbert's program of (c), page 146, the restriction is not only acceptable, but necessary. But the fact that the intuitively significant distinction above cannot be so formulated, reduces the foundational importance of a formalistic analysis, by requirement (ii) of informal rigour.]

Next it is not surprising that there is a certain *asymmetry* between the role of higher order consequence for derivability results (section 1a) and its role in independence results (section 1b). The same is familiar e.g. from recursion theory. Thus to establish negative (i.e., unsolvability) results one will aim in the first place to show recursive unsolvability, while to show solvability one gives a *particular* schema and a *proof* showing that the schema works. (Similarly, cf. end of the introduction; even if a problem *is* recursively solvable, one may wish to explain why it has not been solved: by showing e.g. that there is no schema of a given kind which can be *proved* to work by given methods, or else by showing that calculations are too long.) Curiously enough, this obvious point is sometimes overlooked.

Finally, and this is of course the most direct link between the present section and the main theme of this article, second order decidability of CH [in the example of (a) above] suggests this: new primitive notions, e.g. properties of natural numbers, which are *not* definable in the language of set theory (such as in the footnote on p. 150), may have to be taken seriously to decide CH; for, what is left out when one replaces the second order axiom by the schema, are precisely the properties which are not so definable. But I am sure I don't know: the idea is totally obvious; most people in the field are so accustomed to working with the restricted language that they may simply not succeed in taking other properties seriously; and, finally, compared with specific examples that come to mind, e.g. the footnote on p. 150, the so-called axioms of infinity [1] which are formulated in first order form are more efficient.

2. *Intuitive logical validity, truth in all set-theoretic structures, and formal derivability.* We shall consider formulae α of finite order (α^i denoting formulae of order i), the predicate $Val\,\alpha$ to mean: α is intuitively valid, $V\alpha$: α is valid in all set-theoretic structures,

and $D\alpha$: α is formally derivable by means of some fixed (accepted) set of formal rules.

For reference below. $V\alpha$ is definable in the language of set theory, and for recursive rules $D\alpha$ is definable *uniformly*, i.e. for each $\sigma \geqslant \omega$, the same formula defines D when the variables range over C_σ. Below, we shall also consider $V_C\alpha$: validity in classes (i.e. the universe of the structure is a class and the relations are also classes, or, in the terminology of App. A, set theoretic properties) at least for formulae of first and second order.

What is the relation between Val and V?

(a) *Meaning of Val.* The intuitive meaning of *Val* differs from that of V in one particular: $V\alpha$ (merely) asserts that α is true in all structures in the cumulative hierarchy, i.e., in all sets in the precise sense of *set* above, while $Val\alpha$ asserts that α is true in *all* structures (for an obvious example of the difference, see p. 154). A current view is that the notion of arbitrary structure and hence of intuitive logical validity is so vague that it is absurd to ask for a proof relating it to a precise notion such as V or D, and that the most one can do is to give a kind of plausibility argument.

Let us go back to the fact (which is not in doubt) that one reasons in mathematical practice, using the notion of consequence or of logical consequence, freely and surely, (and, recall p. 145; the 'crises' in the past in classical mathematics by (c), page 145, were not due to lack of precision in the notion of consequence.) Also, it is generally agreed that at the time of Frege who formulated rules for first order logic, Bolzano's set-theoretic definition of consequence had been forgotten (and had to be rediscovered by Tarski); yet one recognised the validity of Frege's rules (D_F). This means that implicitly

$$\forall i \forall \alpha (D_F \alpha^i \rightarrow Val \alpha^i)$$

was accepted, and therefore certainly *Val* was accepted as meaningful.

Next, consider the two alternatives to *Val*. First (e.g. Bourbaki) 'ultimately' inference is nothing else but following formal rules, in other words D is primary (though now D must not be regarded as defined set-theoretically, but combinatorially). This is a specially peculiar idea, because 99 per cent of the readers, and 90 per cent of the writers of Bourbaki, don't have the rules in their heads at all! Nobody would expect a mathematician to work on groups if he did not know the definition of a group. (By section 1b, the notion of set is treated in Bourbaki like *Val*.)

Second, consider the suggestion that 'ultimately' inference is semantical, i.e. V is meant. This too is hardly convincing. Consider a formula α^1 with the binary relation symbol E as single non-logical constant; let α_ε mean that α is true when the quantifiers in α range over all sets and E is replaced by the membership relation. (Note that α_ε is a first order formula of set theory.) Then intuitively one concludes:

If α is logically valid then α_ε, i.e. (in symbols): $Val\alpha \to \alpha_\varepsilon$. But one certainly does not conclude *immediately*: $V\alpha \to \alpha_\varepsilon$; for α_ε requires that α be true in the structure consisting of all sets (with the membership relation); its universe is not a set at all. So $V\alpha$ (α is true in each set-theoretic structure) does not allow us to conclude α_ε 'immediately': this is made precise by means of the results in (b) and (c) below. On the other hand one does accept

$$\forall i \forall \alpha (Val\alpha^i \to V\alpha^i)$$

the moment one takes it for granted that logic *applies* to mathematical structures.

Nobody will deny that one knows more about Val after one has established its relations with V and D; but that doesn't mean that Val was vague before.

In fact we have the *theorem*:

For $i=1$, given the two accepted properties of *Val* above,

$$\forall \alpha^1 (Val\alpha \leftrightarrow V\alpha) \text{ and } \forall \alpha^1 (Val\alpha \leftrightarrow D\alpha).$$

The proof uses Gödel's completeness theorem: $\forall \alpha^1 (V\alpha \to D\alpha)$. Combined with $\forall \alpha^i (D\alpha \to Val\, \alpha)$ above, we have $\forall \alpha^1 (V\alpha \leftrightarrow D\alpha)$, and with $\forall i \forall \alpha (Val\, \alpha^i \to V\alpha^i)$ above: $\forall \alpha^1 (V\alpha \leftrightarrow Val\, \alpha)$. Without Gödel's completeness theorem we have from the two accepted properties of *Val*: $\forall \alpha^1 (D\alpha \to V\alpha)$, incidentally a theorem which does not involve the primitive notion *Val* at all.

At least, *Val* is not too vague to permit a *proof* of its equivalence to V for first order α, by use of the properties of *Val* above!

(b) *The relation between* $V\alpha^1$ (α containing a binary E as single non logical constant) *and* α_ε. To discuss this it is convenient to use the theory of explicitly definable properties (App. A: usually called: theory of classes) and the relation

$$Sat(A, B, \alpha)$$

to mean: the property A and the relation B ($\subset A \times A$) satisfy α. We can represent finite sequences of classes $A_1, \ldots, A_p$ by a single class $A = \{\langle n, x\rangle : n \leqslant p \ \&\ x \in A_n\}$. If each A_i is explicitly definable so is A. Now, by standard techniques of forming truth definitions, $Sat(A, B, \alpha)$ is defined by

$$\exists C \Sigma(A, B, C, \alpha)$$

where Σ does not contain class variables [1] other than (the free variables) A, B, C.

Let U stand for the class of all sets, and E for the membership relation restricted to U.

For each particular $\tilde{\alpha}^1$ *we have*: $V\tilde{\alpha}^1 \rightarrow Sat(U, E, \tilde{\alpha})$ *provable in the theory of classes with axiom of infinity, hence* $V\tilde{\alpha}^1 \rightarrow \tilde{\alpha}_\varepsilon$.

Cor. By a well-known result of Novak (see App. A), $V\tilde{\alpha}^1 \rightarrow \tilde{\alpha}_\varepsilon$ is provable in set theory for each formula $\tilde{\alpha}$.

The proof of the theorem uses $\forall\alpha(V\alpha^1 \rightarrow D_G\alpha^1)$ for *cut free* rules (e.g. Gentzen's), and then, for fixed $\tilde{\alpha}$, $D_G\tilde{\alpha} \rightarrow Sat(U, E, \tilde{\alpha})$ by means of a truth definition for *subformulae* of $\tilde{\alpha}$. For the proof of $\forall\alpha(V\alpha^1 \rightarrow D_G\alpha^1)$ one needs of course the axiom of infinity since some α are valid in all *finite* structures without being logically valid.

The machinery needed for this proof certainly justifies the

---

1 The definition has the following *invariance* property. If the set variables in Σ range over a C_σ of the cumulative hierarchy, the classes are objects of $C_{\sigma+1}$, and the particular formula above, for given A and B, defines the same set of α whether C ranges over all of $C_{\sigma+1}$, or only over elements of $C_{\sigma+1}$ explicitly definable from A and B. The corresponding case for *higher* order formulae is quite different.

reservations above against the assumption that we simply *mean* V (i.e. truth in all set theoretic structures in some precise sense of set) when speaking of logical validity. Note incidentally, if we take any suitable *finitely* axiomatised set theory S, there is an $\tilde{\alpha}$ for which $V\tilde{\alpha} \rightarrow \tilde{\alpha}_\varepsilon$ is not provable in S (namely, take for $\tilde{\alpha}$ the negation of the conjunction of the axioms of S, granted that a 'suitable' set theory cannot prove its own consistency, i.e. not $\sim V\tilde{\alpha}$).

The doubts are further confirmed by

(c) $\forall \alpha^1[V\alpha^1 \rightarrow Sat(U, E, \alpha^1)]$ *is not provable in the theory of classes.*

If it were, by the main result of App. A, we should have an explicit set theoretic $\mathsf{F}\alpha^1$ such that

$$\forall \alpha^1[V\alpha \rightarrow \Sigma(U, E, \mathsf{F}\alpha, \alpha)]$$

is provable in the theory of classes. This reduces to a purely set-theoretic formula

$$(*) \qquad \forall \alpha^1[V\alpha \rightarrow \Sigma_1(\mathsf{F}\alpha, \alpha)]$$

which is proved in a finite subsystem S_1 of set theory; regarded as a formal object of predicate logic (with $\in$ replaced by a binary relation symbol E!), let S_1 be σ_1 and let (*), regarded as a formal object, be π. Thus

$$\vdash_{S_1} V(\sigma_1 \rightarrow \neg\, \pi) \rightarrow \Sigma_1[\mathsf{F}(\sigma_1 \rightarrow \neg\, \pi), \sigma_1 \rightarrow \neg\, \pi].$$

But for the particular formula $\sigma_1 \rightarrow \neg\, \pi$, without use of induction, we verify that $\langle U, E\rangle$ satisfies $\sigma_1 \rightarrow \neg\, \pi$, i.e.

$$\vdash_{S_1} \Sigma_1[\mathsf{F}(\sigma_1 \rightarrow \neg\, \pi), \sigma_1 \rightarrow \neg\, \pi] \rightarrow (S_1 \rightarrow \neg\, \forall \alpha[V\alpha \rightarrow \Sigma_1(\mathsf{F}\alpha, \alpha)]).$$

Since $\vdash_{S_1} S_1$, and, by assumption, $\vdash_{S_1}(*)$, we have $\vdash_{S_1} \neg\, V(\sigma_1 \rightarrow \neg\, \pi)$. But this would prove the consistency of S_1 in S_1.

Since *evidently on the intended interpretation of the theory of classes* (explicitly definable properties) (*) *is valid, we have found an instance of ω-incompleteness.* Thus looking at the intuitive relation *Val*, leads one not only to formal proofs as in (a) but also to incompleteness theorems.

(d) All this was for first order formulae. *For higher order formulae we do not have a convincing proof* of e.g. $\forall\alpha^2(V\alpha \leftrightarrow Val\ \alpha)$ though one would expect one. A more specific question can be formulated in terms of the hierarchy of types C_σ. Let V^σ mean: truth in all structures that belong to C_σ.

Then $\forall\alpha^1\forall\sigma>\omega(V^{\omega+1}\alpha \leftrightarrow V^\sigma\alpha)$ (Skolem–Löwenheim theorem). What is the analogue (to ω) for second order formulae? e.g. if $\tilde{\alpha}^2$ is Zermelo's system of axioms, $V^{\omega+\omega+1}(\neg\,\tilde{\alpha}^2)$ is false, $V^{\omega+\omega}(\neg\,\tilde{\alpha}^2)$ is true. This analogue to ω is certainly large. Let α assert of the structure $\langle a, e\rangle$ that (i) it is a C_σ for limit numbers σ, i.e. that $\langle a, e\rangle$ satisfies Zermelo's (second) order axioms, (ii) $\langle a, e\rangle$ contains a measurable cardinal $>\omega$. Here (i) is of second order, and (ii) is of first order relative to (i). If $\beta=(\exists ae)\alpha$, we have $V^\sigma(\neg\,\beta)$ for $\sigma\leqslant$ the first measurable cardinal $\varkappa$, but not $V^\sigma(\neg\,\beta)$ for $\sigma>\varkappa$.

Since we do not even know a reduction analogous to the basic Skolem–Loewenheim theorem, it is perhaps premature to ask for an analogue to $\forall\alpha^1(V^{\omega+1}\alpha \rightarrow D\alpha)$. For instance, a well rounded theory of higher order formulae may be possible only for infinitely long ones. For infinite first order formulae we do know an analogue when $D\alpha$ is replaced by certain *generalized inductive* definitions (cf. ω-rule).

General Conclusion. There is of course nothing new in treating Val as an understood concept; after all Gödel established completeness without having to mention V; he simply used implicitly the obvious $\forall\alpha(Val\ \alpha \rightarrow V^{\omega+1}\alpha)$ and $\forall\alpha(D\alpha \rightarrow Val\ \alpha)$ (incidentally for all i!), and proved $\forall\alpha^1(V^{\omega+1}\alpha \rightarrow D\alpha)$. It seems a good time to examine this *solved* problem carefully because (besides Heyting's rules for intuitionist validity, cf. section 3) we face problems about finitist validity (Val_{Fi}) and predicative validity (Val_P) not unlike those raised by Frege's rules. Thus, as in his case, we have (recursive) rules D_{Fi} and D_P for finitistic and predicative deductions respectively, established by means of autonomous progressions; and then equivalence to Val_{Fi}, resp. Val_P (for the languages considered) is almost as plausible as was $Val \leftrightarrow D$ at the time of Frege. But we have not yet found principles as convincing as those of section 2(a) above to clinch the matter; in fact we do not have an analogue to V.

3. *Brouwer's thinking subject.* The present section considers a striking use of a new primitive notion (which has no place in mathematical practice) to derive a purely mathematical assertion [1]:

$$(*) \sim \forall\alpha[\sim\sim\exists x(\alpha x=0)\supset\exists x(\alpha x=0)],$$

for α ranging over free choice sequences of 0,1.

For general background see the last chapters of Heyting's book on intuitionism and of Kleene's recent monograph [3]. – N.B. Knowledge of these two works is a minimum requirement for discussing profitably the present topic; it would be useful to know in addition the material summarized in section 2 of my recent review article [4]. It is quite unreasonable to want an explanation of intuitionistic notions for the 'man in the street': he simply does not use concepts to which the intuitionistic distinctions apply; if one only does numerical arithmetic (which is decidable) one has no examples of the failure of the law of the excluded middle; and if one only knows arithmetic, one has little chance of grasping distinctions which apply specifically to free choice sequences. It is generally agreed that intuitionistic mathematics is less straightforward than set theory; granted this, one cannot be surprised that what most people know about it, corresponds to little more than Venn diagrams in set theory. These diagrams are an inadequate preparation for discussing really problematic matters in set theory such as axioms of infinity; and (*) is problematic for intuitionistic mathematics.

(a) *Criticisms of Brouwer's argument; a distinction.* It is not necessary to state the argument since it will be given (in modified and formally correct form) in (c) below. Kleene's objection on top of [3], p. 176, to a formal weakness in Brouwer's original version does not apply to (c).

For mathematical practice it would be interesting to have a proof of (*) from as elementary and familiar assumptions as possible. Now, for instance in [4] 2.741, (*) is derived from the assumption

---

1 Important improvements are contained in the *Discussion*. To avoid misunderstandings, I have added references to these points (July 1966).

that all constructive functions are recursive. This assumption is, indeed, in elementary terms, only it is far from plausible, if we understand: constructive, in an intuitionistic sense, not in the sense of: mechanically computable. So, the argument of (c) leaves open whether there is a more elementary derivation. (An elementary derivation, free from doubtful hypotheses, is given in section 2c of my reply to Heyting).

But for truly foundational research it is of special interest to derive a purely mathematical assertion from axioms concerning a specifically intuitionistic notion, here: the thinking subject pursuing mathematics indefinitely (though not necessarily continuously). And, apart from these mathematical consequences, one wants to formulate as fully as possible properties of these basic notions: one learns more about them by getting contradictions (from defective formulations) than by trying to avoid the notions! In Brouwer's own philosophy (or: analysis) of mathematics theorems are supposed to be about mental acts of a thinking subject; more precisely, of a *correctly* thinking subject. Brouwer's views may be wrong or crazy (e.g. self-contradictory), but one will never find out without looking at their more dubious aspects. (This little sermon is beautifully illustrated by Myhill's contribution).

Superficial examination may suggest that the restriction to *correctly* thinking subjects makes the notion of: thinking subject, wholly empty. That this is not so is shown by (c) below: one of the main purposes of the analysis is to restrict the notion of thinking subject so as to eliminate *accidental* psychological elements, yet to exploit essential ones. (Of course, it was not immediately evident that such a compromise can be found in particular, in axiom b (i) below.)

(b) *Axioms.* The basic notion is

$$\textstyle\sum \vdash_m A:$$

the (thinking) subject $\sum$ has evidence for asserting A at stage m. The parameter m will be particularly important for statements A about free choice sequences α, for which, at stage m, only the values $\alpha(0), \ldots, \alpha(m-1)$ are given.

(i) $\sum \vdash_m A$ is decidable for each given $\sum$, m, A;

(ii) $A \supset \forall \sum \sim \sim \exists m(\sum \vdash_m A)$ and $\forall \sum[\exists m(\sum \vdash_m A) \supset A]$

(universality of mathematics).

The logical particles in these axioms are to be interpreted constructively. This is consistent with Heyting's explanation if it is accepted that the (intension of the) range of the variable $\sum$ is *given*, as of other variables, for instance for integers, free choice sequences, etc.

(c) *Deduction of* (*) *from the axioms* (b) *and current intuitionistic axioms.* Let $P(\beta, \sum, \alpha, m)$ stand for:

$$(\alpha m = 0) \leftrightarrow [(\exists x < m)(\beta x \neq 0) \vee \textstyle\sum \vdash_m \forall x(\beta x = 0)].$$

(i) $(\forall \beta \sum \alpha)[\forall m P(\beta, \sum, \alpha, m) \supset \sim \sim \exists n(\alpha n = 0)]$.

Since, if $\forall m P(\beta, \sum, \alpha, m)$ and $\beta n \neq 0$ then $\alpha(n+1) = 0$, $(\exists x < n+1)(\beta x \neq 0)$ being true, we have

$$[\forall m P(\beta, \textstyle\sum, \alpha, m) \;\&\; \forall m(\alpha m \neq 0) \supset \forall n(\beta n = 0)];$$

but, by axiom (ii),

$$\forall n(\beta n = 0) \supset \forall \textstyle\sum' \sim \sim \exists p[\sum' \vdash_p \forall x(\beta x = 0)]$$

and so

$$\forall m P(\beta, \textstyle\sum, \alpha, m) \;\&\; \forall m(\alpha m \neq 0) \supset \sim \sim \exists p(\alpha p = 0),$$

whence

$$\forall m P(\beta, \textstyle\sum, \alpha, m) \supset \sim \sim \exists p(\alpha p = 0).$$

(ii) Assuming (*) we get

$$(\forall \beta \textstyle\sum)\exists m'[(\exists x < m')(\beta x \neq 0) \vee \sum \vdash_{m'} \forall x(\beta x = 0)].$$

For, by the previous step, and (*), we have

$$(\forall \beta \textstyle\sum \alpha)[\forall m P(\beta, \sum, \alpha, m) \supset (\exists m')(\alpha m' = 0)]$$

and so

$$(\forall \beta \textstyle\sum \alpha)[\forall m P(\beta, \sum, \alpha, m) \supset (\exists m')\{(\exists x < m')(\beta x \neq 0) \vee \sum \vdash_{m'} \forall x(\beta x = 0)\}].$$

But, by b(i) (decidability of the basic relation)

$$(\forall \beta \textstyle\sum) \exists \alpha \forall m P(\beta, \textstyle\sum, \alpha, m)$$

and so α drops out, leaving

$$(\forall \beta \textstyle\sum) \exists m' \{(\exists x < m')(\beta x \neq 0) \vee \textstyle\sum \vdash_{m'} \forall x(\beta x = 0)\}.$$

(iii) Now by the second half of b(ii), $\sum$ drops out, leaving

$$\forall \beta [(\exists x < m')(\beta x \neq 0) \vee \forall x(\beta x = 0)].$$

But this contradicts continuity, and so (*) has led to a contradiction, as required. (The last step in this proof needs additional analysis; cf. Myhill's section 6 and section 3 of my reply to him.)

It might be remarked that, since (*) is a negative assertion, one only needs the double negation of the axioms b(i)–(ii).

A reader familiar with the detailed theory of free choice sequences will have noticed that the argument leaves open the following question: Suppose Sa means that the (constructive) function a defines a spread; then (*) is equivalent to

$$(\forall a)[Sa \mathbin{\&} (\forall \alpha \in a) \sim\sim \exists x(\alpha x = 0) \supset (\forall \alpha \in a) \exists x(\alpha x = 0)].$$

The argument above does not exclude the possibility of a purely *mathematical* proof showing how, for each a, from a *mathematical* proof of the premise one gets a *mathematical* proof of the conclusion. (For a much more precise formulation, see the discussion.)

Obviously one would wish to make the new basic notion an object of independent study before judging its value. But, independently of this the deduction illustrates a general point. Intuitionism is, at present, an excellent field for foundational research because its foundations have not been treated much and hence arguments as simple as (c) can settle a formally undecided problem. For the same reason it lends itself much less to technical research of a level comparable with that of classical practice [1].

---

[1] Kleene calls $\forall \alpha [\sim\sim \exists x(\alpha x = 0) \supset \exists x(\alpha x = 0)]$ a generalisation of Markoff's principle, and Heyting said in the discussion that Markoff would formulate the principle only for constructive functions in place of free choice sequences α. While this distinction is certainly valid, it seems to

Appendix A: Axioms for explicitly definable properties

Let $\mathscr{L}$ be a language of first order (without function symbols) whose relation symbols are R_i, R_i having $\varrho(i)$ arguments. For any formula $C(x, y, z)$ with three free variables (later to be interpreted as 'pairing relation') define $C_n(x_1, \ldots, x_n, y)$ as follows:

$$C_2(x_1, x_2, y) \quad \text{is} \quad C(x_1, x_2, y),$$

and

$$C_{n+1}(x_1, \ldots, x_{n+1}, y) \quad \text{is} \quad \exists z[C_n(x_1, \ldots, x_n, z) \,\&\, C_2(z, x_{n+1}, y)].$$

(Later one will interpret $C_n(x_1, \ldots, x_n, y)$ as $y = \langle x_1, \ldots, x_n \rangle$, i.e. ordered n-tuple relation.)

Let $\mathscr{L}_{\mathscr{P}}$ be the *extension* of $\mathscr{L}$ obtained by adding a second sort of variable (capital letters) and the binary relation η, $x\eta X$ to be interpreted as: the object x has the property X.

Basic axioms.

1. $\forall x \exists X \forall y(y\eta X \leftrightarrow y = x)$,
2. $\exists X \forall y[y\eta X \leftrightarrow \exists x C(x, x, y)]$,
3. $\forall X \exists Y \forall y(y\eta X \leftrightarrow \neg\, y\eta X)$,
4. $\forall XY \exists Z \forall y[y\eta Z \leftrightarrow (y\eta X \vee y\eta Y)]$,
5. $\forall X \exists Y \forall y(y\eta Y \leftrightarrow \exists xz[C(y, x, z) \,\&\, z\eta X])$,
6. $\forall X \exists Y \forall y(y\eta Y \leftrightarrow \exists xz[C(x, z, y) \,\&\, z\eta X])$,
7. $\forall X \exists Y \forall y(y\eta Y \leftrightarrow \exists uvz[C(u, v, y) \,\&\, C(v, u, z) \,\&\, z\eta X])$,
8. $\forall X \exists Y \forall y(y\eta Y \leftrightarrow \exists uvwz[C_3(u, v, w, y) \,\&\, C_3(v, w, u, z) \,\&\, z\eta X])$,
9. $\forall X \exists Y \forall y(y\eta Y \leftrightarrow \exists uvwz[C_3(u, v, w, y) \,\&\, C_3(u, w, v, z) \,\&\, z\eta X])$,

10^i. $\exists X \forall y(y\eta X \leftrightarrow \exists x_1 \ldots x_{\varrho(i)}[C_{\varrho(i)}(x_1, \ldots, x_{\varrho(i)}, y) \,\&\, R_i(x_1, \ldots, x_{\varrho(i)})]$.

(2 is subsumed under 10^i if equality is counted among the R_i.) Note that the axioms are relative to the formula C. We call the axioms: $\mathscr{P}_{\mathscr{L}}$ (P for: properties).

---

me too technical: Markoff's (implicit) interpretation of logical connectives is so mechanistic that any similarity to the intended intuitionistic interpretation is purely coincidental. The argument above no more contradicts Markoff's principle as he understands it than it contradicts the usual *classical* reading of $\forall\alpha[\sim\sim\exists x(\alpha x = 0) \supset \exists x(\alpha x = 0)]$. – This much, I believe, is clear; it is not quite so clear that the rules of intuitionistic mathematics are valid at all for Markoff's interpretation (if the latter is made explicit).

Definition. $C(x, y, z)$ is called a *pairing formula* relative to a set $\mathfrak{A}$ of axioms in $\mathscr{L}$ if and only if

$$\mathfrak{A} \vdash \forall xy \exists z C(x, y, z) \ \& \ \forall xx'yy'z[\{C(x, y, z) \ \& \ C(x', y', z)\} \rightarrow \rightarrow (x = x' \ \& \ y = y')].$$

For example, if $\mathscr{L}$ is the language of set theory (Zermelo), and $\mathfrak{A}$ certain elementary axioms, $z = \{\{x\}, \{x, y\}\}$ can be taken for C. Note that, if $\mathfrak{A}$ has finite models, it does not admit a pairing relation.

Theorem. Suppose $\Phi(X_1, \ldots, X_r)$ is a formula of $\mathscr{L}_{\mathscr{P}}$ built up from prime formulae $t\eta X_i$, $(i \leqslant r)$ using the operations of $\mathscr{L}$ *only* (t denotes terms of $\mathscr{L}$). Let $\mathfrak{A}$ be a set of formulae of $\mathscr{L}$ itself. Then $\mathfrak{A}, \mathscr{P}_{\mathscr{L}} \vdash \exists X_1 \ldots X_r \Phi(X_1, \ldots, X_r)$ if and only if there are formulae $F_i(x, u_1, \ldots, u_p)(i \leqslant r)$ of $\mathscr{L}$ such that $\mathfrak{A} \vdash \exists u_1 \ldots u_p \Phi(F_1, \ldots, F_r)$ where $\Phi(F_1, \ldots, F_r)$ is obtained from $\Phi(X_1, \ldots, X_r)$ by replacing $t\eta X_i$ by: $F_i(t, u_1, \ldots, u_p)$; it is supposed that all the variables in F_i are distinct from all the variables in $\Phi(X_1, \ldots, X_r)$ to avoid clashing.

Corollaries. 1. If Φ is a formula of $\mathscr{L}$ and $\mathfrak{A}, \mathscr{P}_{\mathscr{L}} \vdash \Phi$ then also $\mathfrak{A} \vdash \Phi$.

2. Suppose $\Psi(X)$ is built up from prime formulae $t\eta X$ by using the operations of $\mathscr{L}$ only. Then, for Φ in $\mathscr{L}$, $\mathfrak{A}, \forall X \Psi(X) \vdash \Phi$ if and only if there is an F in $\mathscr{L}$ such that $\mathfrak{A} \vdash \forall u_1 \ldots u_p \Psi(F) \rightarrow \Phi$.

Proof of Theorem. Suppose $\mathfrak{A} \vdash \exists u_1 \ldots u_p \Phi(F_1, \ldots, F_r)$. To show that $\mathscr{P}_{\mathscr{L}}, \mathfrak{A} \vdash \exists X_1 \ldots X_r \Phi(X_1, \ldots, X_r)$ it is clearly sufficient to show: $\mathscr{P}_{\mathscr{L}} \vdash \forall u_1 \ldots u_p \exists X_i \forall x[x\eta X_i \leftrightarrow F_i(x, u_1, \ldots, u_p)]$. For this one uses a (metamathematical) induction argument according to the complexity of F_i just as in the proof of Bernays class theorem. A minor difference is that if $\mathfrak{A}$ are the axioms of set theory, $=$ is explicitly definable from $\in$ because of extensionality.

For the converse, one uses the fact that any model

$$\langle A, \bar{R}_1, \ldots, \bar{R}_i, \ldots \rangle$$

of $\mathfrak{A}$ can be *extended* to a model of $\mathfrak{A}$ and $\mathscr{P}_{\mathscr{L}}$, in particular to a minimal model as follows: one takes as the domain of the second sort of variable the subsets of A which are explicitly definable in

$\mathscr{L}$ by means of formulae containing individual constants for elements of A, and for $\bar{\eta}$ the following relation: $\overline{X\eta t}$ always false; for elements a, b of A, $\overline{a\eta b}$ false; and $\overline{a\eta X} \Leftrightarrow \bar{a} \in \bar{X}$.

Now, suppose $\mathfrak{A}, \mathscr{P}_{\mathscr{L}} \vdash \exists X_1 \dots X_r \Phi(X_1, \dots, X_r)$. If $F_1^{(i)}, \dots, F_r^{(i)}$ $(i=1, 2, \dots)$ is a list of all the formulae in $\mathscr{L}$, the set

$$\mathfrak{A} \cup \{\dots, \forall u_1 \dots u_{p(i)} \neg \Phi(F_1^{(i)}, \dots, F_r^{(i)}), \dots\}$$

is inconsistent in each model of $\mathfrak{A}$, because then

$$\exists X_1 \dots X_r \Phi(X_1, \dots, X_r)$$

would be false in the minimal extension above. So, by the finiteness theorem some subset is inconsistent, and so

$$\mathfrak{A} \vdash \exists u_1 \dots u_{p(1)} \Phi(F_1^1, \dots, F_r^1) \vee \dots \vee \exists u_1 \dots u_{p(k)} \Phi(F_1^k, \dots, F_r^k).$$

This disjunction can be *contracted* as follows, when

$$q = \max\,[p(1), \dots, p(k)]:$$

$$\begin{aligned}
&F_i^*(u_1, \dots, u_q, x) \leftrightarrow . \\
&[F_i^1(u_1, \dots, u_{p(1)}, x) \;\&\; \exists u_1 \dots u_{p(1)} \Phi(F_1^1, \dots, F_r^1)] \vee \\
&[F_i^2(u_1, \dots, u_{p(2)}, x) \;\&\; \exists u_1 \dots u_{p(2)} \Phi(F_1^2, \dots, F_r^2) \;\&\; \neg \\
&\qquad\qquad \neg \exists u_1 \dots u_{p(1)} \Phi(F_1^1, \dots, F_r^1)] \vee \dots \\
&[F_i^k(u_1, \dots, u_{p(k)}, x) \;\&\; \neg \exists u_1 \dots u_{p(1)} \Phi(F_1^1, \dots, F_r^1) \;\& \\
&\qquad\qquad \& \dots \& \neg \exists u_1 \dots u_{p(k-1)} \Phi(F_1^{k-1}, \dots, F_r^{k-1})].
\end{aligned}$$

So $\mathfrak{A} \vdash \exists u_1 \dots u_q \Phi(F_1^*, \dots, F_r^*)$.

The same argument shows a little more, namely in the case when $\mathfrak{A}$ does not possess a pairing relation.

Suppose $\mathfrak{A}$ has only infinite models; add to $\mathscr{L}$ a symbol $C(x, y, z)$ with the axioms $\mathfrak{C}$ for a pairing relation, and form $\mathscr{P}_{\mathscr{L}'}$ for the enriched language $\mathscr{L}'$. If Φ is a formula of $\mathscr{L}$ and a consequence of $\mathfrak{A} \cup \mathfrak{C} \cup \mathscr{P}_{\mathscr{L}'}$ it is also a consequence of $\mathfrak{A}$; for any model of $\mathfrak{A}$ can be enriched to one of $\mathfrak{A} \cup \mathfrak{C}$, and then one applies the argument above to $\mathscr{L}'$ instead of $\mathscr{L}$.

Remark. The idea of the proof is absolutely standard, all well-known from the theory of classes. It is only the formulation for arbitrary $\mathscr{L}$ that does not seem to be given in the literature.

The use of a *new* symbol η is analogous to Bernays' formulation of the (oddly named) theory of classes, regarded as a theory of explicitly definable set-theoretic properties. To obtain the same result for Gödel's variant, where $\mathscr{L}$ is the language of set theory, one observes this. $\mathfrak{A} \vdash \forall x \exists y(x \in y)$, for the axioms $\mathfrak{A}$ of Zermelo. Any model of $\mathfrak{A}$ is extended to one of the theory of classes by *identifying* an element a of A with the (explicitly defined) $\{x: x \,\bar{\varepsilon}\, a\}$, so that $\bar{\varepsilon}$ becomes the restriction of $\bar{\eta}$ to $A \times A$. Evidently, since $\mathscr{L}$ contains only a single R_i, 10^i in $\mathscr{P}_{\mathscr{L}}$ collapses to a single formula.

Though the class formalism was originally introduced to deal with purely formal questions of finite axiomatisability, even for technical purposes it is good to ask oneself what classes *are* (informal rigour!).

Appendix B: Standard and nonstandard models

The question which we want to formulate is this: which comes first? So to speak: which is more fundamental? The first point is that, since one speaks of *models* at all, the question is technical and not foundational; cf. beginning of section 1. (This does not deny the interest of nonstandard models unless one believes that *only* ontological questions are interesting!) More precisely, if one thinks of the axioms as *conditions* on mathematical objects, i.e. on the structures which satisfy the axioms considered, these axioms make a selection *among* the basic objects; they do not tell us what the basic objects are. And, if the axioms, in particular the logical symbols, are interpreted 'classically', one assumes that these basic objects support certain operations (complementation, union, projection) satisfying the familiar laws. A precise conception of basic objects which allows this is e.g. the type structure discussed in (b), (page 144).

If this structure *is* the framework in terms of which standard and nonstandard models are discussed, our question can be formulated as a technical question, as in (a) below.

If some other framework is intended, one had better give it a second thought! One coherent (though, by (c) page 151, section 1c, not satisfactory) alternative is a formalist interpretation, for which, trivially, the standard 'model' is certainly not primary since, on this interpretation, all *models* are just manners of speaking. The trouble here is that the 'manners of speaking' actually used in the theory of nonstandard models are not eliminable by use of formalist principles; since eliminability is a defining property of a manner of speaking, this alternative is out.

Another suggestion for a new basic framework may be put as follows. To 'do' arithmetic we do not have to assume the existence of the standard model; *any* (nonstandard) model will do. So, we *reduce existential assumptions* when we do not assume the standard model but only *some* nonstandard models. Is this reduction genuine? There may be doubt about the existence of infinite structures; at least such doubts have been expressed. But, if we have *any* reason for accepting infinite structures at all, isn't this reason furnished by the conception of the standard model? Certainly when so-called strong models of arithmetic are considered, i.e. those that satisfy the same sentences (in a given language) as the standard model, the reduction doesn't strike one as genuine.

Perhaps, as at the end of (c), page 152 on formalisation, one is tempted to give foundational importance to nonstandard models (which *are* important by showing the inadequacy of first order languages) because of their technical uses. It is therefore good to see if these uses support the case, or if nonstandard models fit into ordinary mathematical tradition.

Note first that very often the mathematical properties of a domain D become only graspable when one embeds D in a larger domain D'. Examples: (i) D integers, D' complex plane; use of analytic number theory. (ii) D integers, D' p-adic numbers; use of p-adic analysis. (iii) D surface of a sphere, D' 3-dimensional space; use of 3-dimensional geometry. Non-standard analysis fits in here, and seems obviously superior to other attempts at non-archimedean analysis because it treats integers and real numbers more *symmetrically*; for instance, in a Taylor series $\sum_n a_n x^n$, classical non-archimedean analysis takes a_n in a non-archimedean field K, but the n range over the standard integers, while nonstandard analysis introduces the notion of 'integers of K' and lets n range over them.

Of course, algebraic number theory also considers structures other than the integers which satisfy a given set $\mathfrak{A}$ of properties of the integers, and gets information about them by looking at these other structures. One doesn't conclude from this that there is something dubious about the concept of integer! What is – at least, at present – somewhat exceptional about nonstandard models, is that one uses primarily their existence (i.e. the fact that they satisfy the defining conditions) and not some specific properties beneath the surface. The distinction is somewhat vague, because one has some *constructions* too, e.g. of nonstandard models which satisfy not only given axioms but *exactly* the same formulae, i.e. elementarily equivalent models.

Speaking of existential assumptions, nonstandard models *may* be a more efficient tool for exploiting set-theoretic existential assumptions for getting arithmetic conclusions than, e.g., (i) and (ii) above. Thus, though in (i) one speaks of the complex plane and functions on it, i.e. objects in $C_{\omega+2}$ in Zermelo's hierarchy (section 1), in existing arguments these objects can generally be replaced by the plane of *rational* (complex) numbers and primitive recursive approximations to the functions introduced. This has the effect that the proof uses the comprehension axiom in the very restricted form needed to ensure the existence of primitive recursive classes (i.e. properties in the sense of App. A). In contrast, the moment one talks of a model of all true statements of arithmetic one has a *chance* of needing the comprehension axiom in a stronger form, i.e. ensuring the existence of such a model. Of course, this too may be illusory in practice: cf. (b) below.

To avoid misunderstanding. I am looking here for uses of nonstandard models which require basic axioms *beyond* those employed in current practice. This is in keeping with the present paper: *if* one is going to interpret mathematics set-theoretically or model-theoretically at all, one should try to use the full content of this interpretation by condition (ii) on informal rigour. If one does not want this interpretation the current axioms are themselves problematic and need something like a consistency proof. It is evident that the use of nonstandard models of a theory of types $<\sigma$ can always be formalized in set theory if the existence of σ can be formally proved in set theory.

(a) Which comes first, *within* the basic conceptual framework of classical mathematics: standard or nonstandard models? Note at once that current mathematics is full of related questions; when we look at an axiomatic system we do not merely look for properties common to all models of it (i.e. consequences), but for *principal* structures among them, e.g. minimal ones (the group defined by certain relations, and not only: all groups satisfying them), universal ones, etc.

In foundations too we find good answers to related questions: Which come first: (finite) ordinals or (finite) sets? In the *finite* case, we can define a structure of (hereditarily finite) sets in terms of the $\in$-relation which is *isomorphic* to the structure of the natural numbers with the usual relations of addition, multiplication, exponentiation on it; and conversely, in terms of the latter we can find a structure isomorphic to the structure of sets with the $\in$-relation, e.g. $n \in m$ if $m = 2^{n_1} + 2^{n_2} + \ldots + 2^{n_p}$ (in binary notation) and one $n_i = n$. In the *infinite* case, the relation is asymmetric: in terms of the $\in$-relation we can characterise up to isomorphism (set-theoretic) properties in the sense of App. A: $\langle O, L, \ldots \rangle$ which are the structure of the ordinals with the $<$ relation (and the other usual ordinal-theoretic relations on O), namely, using quantifiers over sets

(i) L is a total ordering of O,
(ii) Every initial segment of $\langle O, L \rangle$ is a well-ordered set,
(iii) To every well-ordered set, i.e. $\langle a, b \rangle$, $b \subset a^2$, b well ordering of a, there is an isomorphic segment of $\langle O, L \rangle$.

But the converse is not true because, essentially, it implies the axiom of constructibility.

In short there is nothing outlandish in our question.

We have the following well-known results:

First, suppose an axiom system (for arithmetic, whose set of Gödel numbers) is *definable* in the standard model, by means of the language considered, i.e. definable in the *structure* consisting of the natural numbers as universe and the relations denoted by the predicate symbols of the language. Then a *nonstandard model* is definable in the standard model (by means of the same language).

In contrast, for any axiom system $\mathfrak{A}$ containing the schema of induction in the language considered and for *every* nonstandard model M of $\mathfrak{A}$, the class of standard integers of M is *not* definable in M. (In fact, under simple conditions no structure isomorphic to the standard model is definable in M.)

Certainly, it is not claimed that the definability criterion above formulates fully the question posed above. But it seems to be at least a sane step towards taking it seriously instead of leaving it at an empty level (cf. (b) page 143).

(b) Axioms that follow from the existence of nonstandard models which satisfy exactly the same (first order) formulae as the standard model. Let $\mathfrak{A}$ be the set of axioms of first order classical arithmetic, and let $R(x)$ be a monadic predicate not in $\mathfrak{A}$. With each closed formula F *not* containing R we associate F_R obtained by restricting all variables in F to R. Then $\mathfrak{A}^*$ consisting of $\mathfrak{A}$, $F \leftrightarrow F_R$ (for all F above), $\forall x \forall y[\{R(x) \;\&\; \neg R(y)\} \rightarrow x<y]$ (where $<$ is a definition of order in $\mathfrak{A}$) is also consistent.

A. Robinson asked: Is $\mathfrak{A}^*$ *a conservative extension of* $\mathfrak{A}$? (If not, the mere existence of a nonstandard model would give information about the standard model.)

We shall answer a *stronger* question. We only assume that $\mathfrak{A}$ is consistent (not necessarily satisfied by the standard model) and includes the schema of induction and we show that $\mathfrak{A}^*$ *is* conservative.

Note that the intended interpretation is that the range of R is the standard model, and that 'really' not $\mathfrak{A}$ is assumed, but $\{F_R\colon F \in \mathfrak{A}\}$. The last axiom above means that the nonstandard objects are beyond the standard ones.

It is sufficient to show this. For *every denumerable* model $M = \langle A, \oplus, \odot\rangle$ of $\mathfrak{A}$ there is an *end model* $\langle A', \oplus', \odot'\rangle$, i.e. additional elements coming after those of A ($A \subset A'$, $\oplus$ and $\odot$ are the restrictions of $\oplus'$, $\odot'$ to A, for every element $a \in A$, $b \in A' - A$ there is a c, $\in A' - A$, $b = a \oplus' c$), which satisfies the same first order sentences as M. Then, if R is interpreted by A the new axioms are satisfied, i.e. $\langle A', A, \oplus', \odot'\rangle$ satisfies the new axioms.

For, as in the proof of App. A, if F is a consequence of $\mathfrak{A}^*$ and not of $\mathfrak{A}$, take a model M which satisfies $\neg F$. It could not be extended to a model M' as above.

The definition of M' follows closely Skolem's original construction in [7] who starts with the *standard* model M (except that here we have to distinguish between variables t over A and variables n over standard integers; in particular, all arguments of the function g in [7] will be standard integers). For details (and an extension to non denumerable models) see [5].

NB. The proof uses the fact that induction is satisfied in the form:

$$\forall x_1 \dots x_k[\{F(0, x_1, \dots, x_k) \mathbin{\&} \forall y[F(y, x_1, \dots, x_k) \to \\ \to F(y+1, x_1, \dots, x_k)]\} \to \forall y F(y, x_1, \dots, x_k)],$$

i.e. that every set defined in a model of $\mathfrak{A}$ by means of a formula F containing constants for elements of the model, has a first element (the contrapositive of induction). This is a consequence of induction without parameters, i.e. with formulae F without $x_1, \dots, x_k$, namely induction applied to $G(z)$:

$$\forall x_1 \dots x_k[\{F(0, x_1, \dots, x_k) \mathbin{\&} \forall y[F(y, x_1, \dots, x_k) \to \\ \to F(y+1, x_1, \dots, x_k)]\} \to F(z, x_1, \dots, x_k)].$$

Clearly, $G(0)$, $\forall y[G(y) \to G(y+1)]$ can be proved, and $\forall z G(z)$ is equivalent to the axiom above.

Remarks (i). The *obvious* proof of the consistency of $\mathfrak{A}^*$ in Robinson's original question uses, of course, the existence of a nonstandard model of all true sentences of arithmetic. For then $F \leftrightarrow F_R$ is automatically satisfied for all F. The existence of a nonstandard model of, say, usual first order classical arithmetic would not in general ensure $F \leftrightarrow F_R$ for *all* F, but only for the axioms $\mathfrak{A}$. As mentioned above the existence of such a 'strong' nonstandard model is proved by use of a fairly strong form of the comprehension axiom. So one might have expected that the axioms $\mathfrak{A}^*$ which are *informally* derived from its existence, are stronger than $\mathfrak{A}$. It turns out that this is not so (as in the traditional applications of analytic number theory).

(ii). A. Robinson pointed out to me (July 1966) that $\mathfrak{A}^*$ might be further strengthened by adding the following schema for *all* formulae F in the extended notation:
From $F0$ and $\forall x[(Rx \mathbin{\&} Fx \to F(x+1)]$ infer $\forall x(Rx \to Fx)$.

H. Friedman has shown that this extension is also conservative (over classical first order arithmetic).

These isolated questions derive special interest from the following *general problem* which is still unsolved. *To find an axiomatic system in which current non-standard arithmetic and analysis can be formalized.* The systems mentioned above allow one to speak of infinite integers (and infinitesimals), namely those satisfying $\neg$R. But there is no evidence as yet that (the second order analogues of) these particular systems are adequate to existing informal uses of non-standard analysis. The discovery of adequate systems is needed if one wants to formulate precisely and justify the kinds of *rules of operation with infinitesimals* which early analysts like Leibniz envisaged. (The *historical* question of what they had in mind seems very difficult: Leibniz himself has written masses of notes that are still unpublished and that would have to be considered. Of course, Robinson's selection of quotations makes entertaining reading, independently of their historical or philosophical relevance.

References

[1] K. Gödel, 'Remarks before the Princeton Bicentennial Conference on problems in mathematics', *The Undecidable*, (ed. M. Davis), New York (1965), pp. 84–88.

[2] W. Hanf and D. Scott, 'Classifying inaccessible cardinals', *Notices Amer. Math. Soc.* 8 (1961), p. 445.

[3] S. C. Kleene and R. Vesley, '*The foundations of intuitionistic mathematics*', Amsterdam (1965).

[4] G. Kreisel, 'Mathematical Logic', *Lectures on modern mathematics*, (ed. Saaty), vol. 3 (1965), pp. 95–195.

[5] R. MacDowell and E. Specker, 'Modelle der Arithmetik', *Infinitistic Methods*, Warsaw (1961), pp. 257–263.

[6] H. Rasiowa and R. Sikorski, '*The mathematics of metamathematics*', Warsaw (1963).

[7] Th. Skolem, 'Peano's axioms and models of arithmetic', *Mathematical interpretation of formal systems*, (eds. Skolem *et al.*), Amsterdam (1955), pp. 1–14.

DISCUSSION

Y. Bar-Hillel: *Obtaining axioms by reflection.*

It is quite customary for axiomatizers to express themselves as if they got their axioms by reflection on certain notions. Hao Wang, for example, mentions in one of his papers how Dedekind described, in a letter to a colleague, the way in which he arrived at his axioms for arithmetic through reflecting on the notion of natural number. Though I have no intention of disparaging the heuristic value of this procedure, I think that there are definite limitations to it. I wonder whether reflection on the notion of set will really help in deciding which formulation of, say, the axiom of comprehension to accept, or whether to accept an axiom of grounding at all. Finally, I cannot see how reflection will direct us to develop a set theory with *Urelemente* (individuals which are not sets) or rather one whose ontology – in Quine's sense – comprises sets only.

Even the *heuristic value* of reflection can be impaired by taking the expression 'reflecting about . . .' too seriously. I would certainly object against taking too seriously the picture, or mythos – in another Quinean phrase – that somewhere there are certain mathematical entities around whose exact nature is somehow veiled to a normal mortal and which reveal themselves only to those who know how to make good use of their reflective capacities.

It would also be of help to some of us if we could understand your notions of *informal rigour* vs. *formal rigour* as being closely similar to Carnap's pair of *clarification of the explicandum* vs. *providing the explicatum*. This identification would be of particular importance if I am correct in assuming that you intend your pair of notions to be used not only in the philosophy of mathematics but in the philosophy of science in general.

A. Heyting: *Informal rigour and intuitionism.*

For me the most interesting part of Kreisel's lecture was his formalization of the introduction of a subject in intuitionistic set theory. I am not quite happy with that introduction. In intuitionistic mathematics the mathematical construction itself is relevant; the construction is objective in this sense that it is irrelevant which subject makes the construction. Would it not be better, from the point of view of informal rigour which he is taking here, to start from what intuitionists really do, that is, to formalize the way in which they make their constructions, without reference to the subject? Kreisel has in an earlier paper made an attempt in this direction and I have the impression that this was more in agreement with the point of view which he defended today.

It is true that Brouwer in his lecture ('Consciousness, philosophy and mathematics', *Proc. 10th International Congress of Philosophy*, Amsterdam 1948, pp. 1235–1249) introduced an entirely new idea for which the subject is of essential importance. As I said, I feel that it is still questionable whether it is possible and whether it is good to introduce this idea in mathematics. But I agree that Kreisel's formalization may be a means of settling this question, in other words, to decide whether Brouwer's notion can be taken seriously, whether a definite meaning can be given to his way of introducing the subject. Van Dantzig has tried another formalization of Brouwer's reasoning (D. van Dantzig, 'Comments on Brouwer's theorem on essentially-negative predicates', *Proc. Akad. Amsterdam* **52** (1949), pp. 949–957 = *Indagationes math.* **11**, pp. 347–355), but he did not succeed in rendering faithfully Brouwer's thought. I am not quite convinced that Kreisel has succeeded better, but this can only be seen after he has further developed his system.

In any case the method in question is not central in intuitionistic mathematics. It can only be applied to show that certain propositions of which nobody believed that they could be true, are actually false.

J. MYHILL*: *Remarks on continuity and the thinking subject.*

(1) The axioms for the thinking subject are *formally inconsistent* with the axiom $(\forall\alpha)(\exists\beta)A(\alpha,\beta) \rightarrow \beta$ depends continuously on α *even when* $A(\alpha, \beta)$ *is purely extensional.* (Proof: let $\beta(n)$ be 0 until α is known to be rational and 1 thereafter: take $A(\alpha,\beta)$ to be

* Professor Myhill was unable to attend the Conference in person.

$\{[(\forall x)(\beta(x)=0) \leftrightarrow \neg\, (\alpha \text{ is rational})] \,\&\, [(\exists x)(\beta(x) \neq 0) \rightarrow \alpha \text{ rational}]\}$; this is extensional but β cannot be chosen to be a continuous function of α.)

(2) On the other hand β cannot be chosen to depend extensionally on α either. So one has to distinguish between extensionality of A and extensionality of the functional which gives β when applied to α).

(3) In his proof of e.g.

$$(**) \qquad \neg(\forall \alpha\beta)[\neg(\alpha=\beta) \rightarrow \alpha \# \beta]$$

Brouwer does in fact *first* construct a discontinuous functional of the kind just mentioned and *then* appeal to continuity via the fan-theorem. How is this discrepancy to be resolved? By what right does he accept the $(\forall\alpha)(\exists x)$-continuity axiom but reject the $(\forall\alpha)(\exists\beta)$-one? Note that in *both* cases the formula following the prefix is extensional and mathematical, while in *neither* case does he suppose it to have been *proved* mathematically. At first sight this seems purely arbitrary; but see (6) below.

(4) (1) above shows that we cannot consistently add the axioms for the thinking subject to either of the usual systems of intuitionistic analysis (Kreisel's or Kleene's) even if we restrict $A(\alpha, \beta)$ in the continuity-axiom to be extensional.

(5) How to formalize intuitionistic analysis after these realizations? *One* respectable course is to replace the $(\forall\alpha)(\exists\beta)$-continuity axiom by Kripke's schema

$$(\exists\beta)\begin{bmatrix}(\forall x)\,(\beta(x)=0) \leftrightarrow \neg A \\ \& \\ (\exists x)\,(\beta(x)\neq 0) \rightarrow A\end{bmatrix}$$

(notice that the $(\forall\alpha)(\exists!\beta)$-continuity axiom is forthcoming as a theorem and can replace the $(\forall\alpha)(\exists\beta)$-axiom in applications); alternatively one can replace $(\forall\alpha)(\exists\beta)$-continuity by the axioms for $\sum$ and insert caveats of extensionality in various other axioms.

(6) Still this seems arbitrary unless we make some further analysis. The following is a first step in this direction. Analyze an (ordinary) free choice sequence (of natural numbers) as an

absolutely free choice sequence of *pairs* $\langle \alpha(i), S_i \rangle$ where $\tilde{\alpha}_i \in S_i$ and where $\{S_i\}$ is a monotone decreasing sequence of spreads. Postulate continuity in the form that if $(\forall \alpha)(\exists \beta)A(\alpha, \beta)$ then each stage of β depends on a finite number of *stages* $\langle \alpha(i), S_i \rangle$ rather than merely on a finite number of *values* $\alpha(i)$. *Now one gets as a theorem that for extensional A the* $(\forall \alpha)(\exists x)$*-continuity axiom holds but* (with suitable additional axioms for $\langle \alpha(i), S_i \rangle$) *the* $(\forall \alpha)(\exists \beta)$*-axiom does not.* (Perhaps this construction (whose details I plan to publish shortly) can be used to demonstrate the independence of the $(\forall \alpha)(\exists \beta)$-axiom from the $(\forall \alpha)(\exists x)$-one.) It seems likely that my conception of free-choice sequence, which resolves this anomaly in Brouwer so neatly, is slightly different from Kreisel's; be that as it may, *the whole dialectic of this chapter in mathematical philosophy is a delightful example of how our formalizations correct our intuitions while our intuitions shape our formalizations.*

(7) None of the *crucial* deductions in this whole history (the derivation of Kripke's schema from the axioms for $\sum$, or of Kreisel's (*) or Brouwer's (**) or the 'negation of Church's thesis' from Kripke's schema, or my derivation of the contradiction, or my derivation of $(\forall \alpha)(\exists x)$-continuity from the axioms for 'stages') is beyond the ability of a very average undergraduate logic student, *once he is told what to prove.* But then, neither was the deduction of Russell's paradox: it may be that to the next generation the idea of a free-choice sequence will be as transparent as the (iterative) conception of set is today (no more and no less!). The analogy bears brooding upon.

G. KREISEL: *Reply.*

(a) *Reply to Professor Bar-Hillel.* Professor Bar-Hillel's comments do not go into any one of the specific examples of informal rigour given in my paper; in fact, the comprehension axiom is discussed explicitly on pp. 143–145, and the perfectly straightforward case of the axiom of regularity (grounding) on page 147 Thus his comments could just as well, or, perhaps, only, be made by someone who had heard the title of the paper without the substance. But even so, his comments seem to me valuable as a clear and forceful statement of the kind of view, incidentally quite widespread, which the paper criticizes, i.e., the view which tells us not to take seriously the traditional question whether an axiomatic characterization of an informal concept is *correct* because (on this view) informal concepts are by their nature too imprecise for this to have an answer.

I should have thought that section 2 (pp. 152–157) of the paper spells out sufficiently the *proof* of equivalence in the case of first order formulae, between the informal notion of logical validity and the formal notion of derivability (and set-theoretic validity), to show that the view just mentioned is false. Of course, to see this, one has at least to glance at the proof! Perhaps the worst effect of the current, allegedly empirical minded, view is that it persuades people *a priori* that there *can* be no proof, and so they don't look at it.

Concerning the equation

$$\frac{\text{clarification of the explicandum}}{\text{providing the explicatum}} = \frac{\text{informal rigour}}{\text{formal rigour}}$$

two things are to be said. First, strictly speaking, the equation does not hold because Carnap certainly denies the possibility of informal *rigour* or *proof*; he would not accept the problem of finding the *correct* explicatum and proving it, but speaks of 'replacing' the prescientific explicandum by an 'adequate' explicatum. Carnap does not reject the possibility of proof outright, but feels convinced of the impossibility or fruitlessness of such a proof as a result of his experience. The examples of the paper are intended to remind us of fruitful cases. Second, one may well ask whether Carnap's theory of philosophical analysis fits in with his practice, in particular, whether, confronted with a particular question on say inductive inference, he doesn't treat the explicandum as a precisely understandable notion with a unique well determined analysis (possibly after having made a few basic informal distinctions, as were needed e.g. in the case of the notion of *set*). So one gets the impression that a certain theory of the nature of philosophical questions prevents him from saying that he is finding a correct definition of the 'explicandum'. This would account for the heavy neologisms which are intended to show that what he is doing is different from traditional aims. This is my impression; I hesitate to go farther than this in the case of Carnap who would certainly weigh his words.

I should not be surprised if informally rigorous analysis of notions, perhaps even of notions from traditional epistemology, should continue to be useful – not only in the philosophy of science, but in science itself. This would be nothing else but the continuation of such analyses as the *derivation* of the equations of motion for a perfect or for a viscous fluid: the equations were found by analyzing certain qualitative (pre-scientific?) notions. It is of course not claimed that *all* traditional notions stand up to rigorous tests, any more than all modern notions or questions do. I should suggest in passing that one had better look at successes rather than at 'limitations' if one wants to get a good idea of the heuristic value of anything.

Finally it may be pertinent (though, I am afraid, also impertinent) to turn the tables and mention some practical consequences of a positivistic philosophy of science; more precisely, how the matter looks to me, an uninformed outsider (who knows the titles without the contents of some papers), in the case of, say, the science of linguistics. First, there is a psychological consequence of all this business about 'replacing' and 'clarifying' instead of: establishing. Isn't it only too tempting to take 'seriously' quite *ad hoc* notions concerning a tiny fragment of experience provided these notions are formulated within accepted terminology? Poor as the new notions may be, they are labelled 'scientific' and replace something pre-scientific: so they can't be worse than the latter. (One looks upon these 'scientific' notions as Dr. Johnson allegedly felt about a badly dancing dog: the point is not whether the dog dances well, but that it dances at all.) Second, more objectively, it is of course clear that for certain specific problems, like mechanical translation, success depends on finding a theory of certain phenomena in mechanistic terms, i.e. replacing the informal notions which we normally use, by mechanically defined ones. Doesn't positivistic philosophy give one a false estimate of the order of difficulty of such problems? Specifically, in the developed sciences one establishes laws obeyed by certain (material or abstract) objects: at least 'it is quite customary for axiomatizers or physicists to express themselves as if . . .'. Only at a *highly* developed stage does one try to account for methods of perception or proof, for optical illusions and failures of (restricted) methods of proof. Even if it is sometimes possible, as in finitist proof theory, to eliminate all reference to the abstract objects it is a fact that the abstract theories were found first. Is it not likely that a linguist who takes positivistic views 'too seriously' will overlook the importance of the abstract element in the successful sciences and will be simply overconfident about the practical possibilities of working in the 'formal mode'? in particular, of doing this without analyzing *seriously* the relations between words which 'normal mortals' would formulate in terms of abstract concepts.

In conclusion let me stress once more (p. 141 of my paper) the conflict between positivistic views and intellectual experience. In connection with such projects as mechanical translation the views described seem to me to lead to uncritical estimates (underestimation of what has to be done and overestimation of what has been done): in most cases I am afraid these are genuine errors, not pretences for ulterior motives. And even if this impression is uninformed, the following is clear: the views lead to blatantly absurd descriptions of actual research! What also can make Bar-Hillel suggest that it is extraordinary for normal mortals to use their reflective capacities? Maybe some of us don't use them very well, but do we use our other capacities so much better?

If I were really convinced that reflection is extraordinary or illusory I should certainly not choose philosophy as a profession; or, having chosen it, I'd get out fast.

(b) *Reply to Professor Heyting.* 1. There are two minor, quite general, points implied by Heyting's intervention with which I do not agree.

(i) A method, or more precisely a notion (here: of the 'thinking subject'), is marginal if it can only be used to analyse what we already know. I am referring to the last paragraph of Heyting's text. I think one major aim of foundations is to extend the range of *theoretical* understanding, and this includes analysing what we already know. But Heyting may well be right in saying that the particular notion in question will never be important *within* intuitionistic mathematics.

(ii) A major aim of the foundations of intuitionistic mathematics is to render faithfully Brouwer's thought. Here I should appeal to the very objectivity (i.e. irrelevance of which subject does mathematics) stressed by Heyting himself. Certainly mathematicians might not have paid attention to intuitionistic concepts for quite a long time, had Brouwer not drawn attention to this area. But once one has come to look *at* this area, and not past it, one doesn't really depend all that much on an outside lead. Specifically, in the present context, one doesn't want to be too faithful to Brouwer because his argument, as it stands, is weak; cf. Kleene's criticism cited in the lecture. In fact, Myhill's analysis suggests that there may have been a definite error, namely a failure to distinguish between extensional and intensional operations on free choice sequences. If anything, I myself was perhaps a little *too* faithful to Brouwer! I was indeed struck by the distinction which I lamely characterized as a difference between mathematical constructions on free choice sequences and those involving empirical concepts. But the sharper and better formulation is due to Myhill who also realized that, in view of the distinction, a new proof of $\sim \forall \alpha[\exists x(\alpha x = 0) \vee \sim \exists x(\alpha x = 0)]$, used in my derivation, is necessary.

2. I shall now try to consider specific points.

(a) First of all, very little of the 'thinking subject' is used in the derivation. Instead of writing $\sum \vdash_n A$, I could write $\sum_n \vdash A$ and read it as: the nth proof establishes A. In other words, the essential point would not be the individual subject, but the idea of proofs arranged in an ω-order (each proof of course being a mental, not necessarily finite, object on the intuitionistic conception). The idea is that one would not make use of any empirical information about the order in which people come to think of proofs. Also, the sequence $\sum_n$ is not itself considered to be given by a rule.

It still seems to me that, if one wants to discuss *intensional* operations, and assertions, the relation $\sum_n \vdash A$ gives illuminating and manageable examples.

(b) The earlier paper on constructions which Heyting refers to is probably in the Stanford Congress volume (1960). A more polished exposition is on pp. 122–129 in Vol. 3 of: *Lectures on Modern Mathematics* (1965). There is no contradiction between the present formulation and the former. All I am doing is *to enrich the earlier formulation by the addition of an ω-order on proofs*, and certain (weak) properties of this order.

In view of the obviously different kind of evidence from which these axioms are derived, it is of considerable interest to distinguish theorems derived by use of these axioms.

(c) I can, in fact, now (1966) derive the principal result

$$(*) \qquad \sim \forall\alpha[\sim\sim\exists x(\alpha x = 0) \supset \exists x(\alpha x = 0)]$$

without use of the thinking subject, even for the extensional interpretation of quantifiers for free choice sequences. (For a discussion of the latter see my reply to Myhill; more formally, the derivation is in terms of the familiar theory of absolutely free choice sequences.)

Note first that for absolutely free choice sequences α, (*) is already known, JSL 23 (1958) p. 382, Remark 7.4, since

$$\forall\alpha \sim\sim \exists x(\alpha x = 0) \quad \text{and} \quad \sim \forall\alpha\exists x(\alpha x = 0).$$

Brouwer's free choice sequences, i.e., those used for analysis, are not absolutely free choice sequences since for the latter $\forall x\exists y A(x, y) \supset \exists\alpha\forall x A(x, \alpha x)$ is in general false. Kripke has pointed out to me that Brouwer himself was aware of the notion of absolutely free choice sequence, *South African J. of Science* **49** (1952) pp. 139–146, particularly the footnote on p. 142; namely a choice sequence where one makes restrictions on future restrictions of freedom. One of the reasons he gives against admitting them is that they would 'endanger the simplicity' . . . 'of further developments'. It appears that they do the opposite!

Consider a Brouwer free choice sequence to be a *pair* of objects, a constructive function a and a totally unrestricted α^*. (In my original paper in JSL 23, I introduced free choice sequences as pairs. The use in the present connection was stimulated by discussions with Gödel.) Here a is to *code* a spread, where a represents the pair of functions a_1, a_2 defined on finite sequences $\mathfrak{a}$ taken from a domain D, a_1 having values 0 or 1 (a characteristic function) and a_2 having values in D. Here a_1 is supposed to be the characteristic function of a spread, with $a(< >)=1$, $[\mathfrak{a}' \subset \mathfrak{a}$ & $a_1(\mathfrak{a})= =1] \supset a_1(\mathfrak{a}')=1$ (tree form); also $a(\mathfrak{a})=1 \to \exists x[a(\mathfrak{a} * x)=1]$, and the idea is that $a_2(\mathfrak{a})$ can be taken for x, i.e.

$$a(\mathfrak{a})=1 \to a(\mathfrak{a} * a_2(\mathfrak{a}))=1.$$

These are conditions C.

For simplicity we shall assign to *every* constructive function $a=(a_1, a_2)$ a spread as follows.

If for all $\mathfrak{a}' \subset \mathfrak{a}$, the conditions C above are satisfied and $a(\mathfrak{a})=1$ then $\mathfrak{a}$ belongs to the new spread.

If not, we take the last $\mathfrak{a}'_0$ for which the conditions are not satisfied and decide that the *only* path in the new spread that extends $\mathfrak{a}'_0$ is simply $\mathfrak{a}'_0, a'_1, a'_2, \ldots$ where $a'_1=a_2(\mathfrak{a}'_0)$ and

$$a_{n+1}=a_2(\mathfrak{a}_0 * a'_1 * \ldots * a'_n).$$

To a totally unrestricted α^*, given up to stage n, i.e. for given $\bar{\alpha}^*n$, we assign the sequence $\bar{\alpha}^*n$ if all $\mathfrak{a}' \subset \bar{\alpha}^*n$ satisfy C; if not, we take the largest $n_0<n$ such that all $\mathfrak{a}' \subset \bar{\alpha}^*n_0$ satisfy C and we assign to $\bar{\alpha}^*n$ the unique extension of $\bar{\alpha}^*n_0$ of length n.

Thus to every α^* we assign a path in the new spread (associated with a) in such a way that (i) if a itself satisfies C then it is identical with the new spread and (ii) the first n steps are decided by $\bar{\alpha}^*n$ and the values of a_1 and a_2 on sequences of length $\leqslant n$. Denote this path by $\mu_a(\bar{\alpha}^*n, m)$ for $m<n$.

Note that μ is evidently extensional in α^* and a for fixed n and m. It is *evidently* extensional in α^* even where α^* occurs in quantified contexts, but possibly intensional in a.

A Brouwer free choice sequence α is then regarded as a pair (a, α^*), and αt is replaced by $\mu_a[\bar{\alpha}^*(t+1), t]$. (*) becomes

$$\sim \forall a \forall \alpha^* [\sim \sim \exists x(\mu_a[\bar{\alpha}^*(x+1), x]=0) \supset \exists x(\mu_a[\bar{\alpha}^*(x+1), x]=0)].$$

Taking for a, say, $a_1(\mathfrak{a})=1$ (universal spread), $a_2(\mathfrak{a})=$arbitrary, e.g. a fixed value, $\mu_a[\bar{\alpha}^*(x+1), x]=\alpha^*x$ and we already know $\forall \alpha^* \sim \sim \exists x(\alpha^*x=0)$ but $\sim \forall \alpha^* \exists x(\alpha^*x=0)$.

(Note that this proof avoids the tiresome syntactic arguments needed to establish the formal independence of $\sim(*)$. Instead, one has to verify that under the translation the axioms of the formal systems considered are provable theorems of a suitably extended theory of absolutely free choice sequences and constructive functions.)

Let $A_n(m)$ be an enumeration of the primitive recursive predicates of one variable m. The proof above establishes

$$\sim \forall n \forall \alpha [\sim \sim \exists x A_n(\bar{\alpha} x) \supset \exists x A_n(\bar{\alpha} x)],$$

and hence

$$\sim \forall n \forall a [\forall \alpha^* \sim \sim \exists x A_n\{\mu_a{}^*(\bar{\alpha}^* x)\} \supset \forall \alpha^* \exists x A_n\{\mu_a{}^*(\bar{\alpha}^* x)\}],$$

where

$$\mu_a{}^*(\bar{\alpha}^* x) = \langle \mu_a(\bar{\alpha}^* 1, 0), \mu_a(\bar{\alpha}^* 2, 1), \ldots, \mu_a(\bar{\alpha}(x+1), x)\rangle.$$

Note however that this is not yet sufficient to establish incompleteness of Heyting's predicate calculus *without some assumption on constructive functions* (such as Church's thesis). For, to apply the argument of JSL **27** (1962) 139–158, one has to find for each a, a formula of predicate logic whose validity is equivalent to $\forall \alpha^* \sim \sim \exists x A_n\{\mu_a{}^*(\bar{\alpha}^* x)\}$, and for this one needs an assumption on possible a.

(c) *Reply to Professor Myhill.* Myhill's very interesting contribution concerns three separate matters which will be considered in decreasing order of importance.

1. *Axiomatisation of intuitionistic analysis.* Let me go back to a *distinction* which is extremely familiar in informal interpretations of intuitionistic systems, but has not previously affected the 'practice' of intuitionistic analysis, namely the distinction between extensional and intensional (i.e. non-extensional) functionals. Since quantifiers are intuitionistically interpreted in terms of functionals (see e.g. [4] 2.33) this distinction affects the interpretation of $\forall \alpha \exists \beta A(\alpha, \beta)$ in Myhill's section 1.

The laws of intuitionistic logic and the axioms of intuitionistic analysis as set up by Kleene or myself happen to be satisfied if $\forall \alpha \exists \beta$ *is interpreted to mean*: there is a function β of α depending extensionally on α.

On the other hand Myhill constructs an example, in section 1, for which there is an evident method, *by appeal to the thinking subject* (or to the ordering of proofs in 2(a) of my reply to Heyting), for associating a β to α such that $A(\alpha, \beta)$, but *no method that depends extensionally on* α.

Technically, a particularly interesting feature of the example is that $A(\alpha, \beta)$ itself depends extensionally on α and β, as Myhill points out. It should be remarked that here we have an interesting difference from the corresponding familiar situation for *continuity*: if $A(\alpha, \beta)$ depends continuously on α and β and if to each α there is a β then β can be chosen continuously in α. (If A does not depend continuously, then, for α and β of *higher* type, we have the well known *failure of the axiom of choice applied to continuous operations*, [4], section 2.43, p. 133.) In fact the continuity axioms may be regarded as a *problematic* version of the axiom of choice in intuitionistic mathematics. For,

$$\forall\alpha\exists\beta A(\alpha, \beta) \supset \exists F\forall\alpha A(\alpha, F\alpha)$$

is wholly unproblematic (for the most general meaning of the existential quantifier) *if* F is taken to range over all, possibly non-extensional, functions F. It is for this most general meaning of the quantifier that the laws of intuitionistic logic are obvious. If one decides to mean by $\forall\alpha\exists\beta$: $\exists F\forall\alpha$ with F extensional, one has to verify the laws (and additional axioms). Granted that extensional functionals are continuous,

$$\forall\alpha\exists!\,\beta A(\alpha, \beta) \supset \exists F\forall\alpha A(\alpha, F\alpha)$$

holds for continuous F provided *uniqueness* is interpreted by

$$[A(\alpha, \beta)\ \&\ A(\alpha', \beta')\ \&\ \forall x(\alpha x = \alpha' x)] \supset \forall x(\beta x = \beta' x)$$

but, I think, not if $\forall x(\alpha x = \alpha' x)$ and $\forall x(\beta x = \beta' x)$ are replaced by $\alpha \equiv \alpha'$, $\beta \equiv \beta'$ respectively, $\equiv$ denoting definitional equality.

By use of Myhill's analysis one can now discuss Heyting's reservations about introducing the thinking subject more sharply. On the one hand it is clear that the latter notion is pointless if at the same time one were to interpret $\forall\alpha\exists\beta$ extensionally (the axioms for $\sum$ which are used – and are obvious for the intensional interpretation – become false). On the other hand the actual practice of intuitionistic *mathematics* seems most elegantly formulated by restricting oneself to extensional operations. So there may be a *practical* conflict here.

On the other hand, for foundations it is evident that intensional operations are fundamental, by the principle stressed throughout my paper: *extensional operations can be defined in terms of intensional ones, but not conversely.*

I have little doubt that, *once one realises the significance of the distinction in the context of analysis* (as Myhill has done) a good axiomatisation will come out without too much trouble. It would be premature to go into this matter in detail before Myhill's own work (to which he refers in his contribution) is published. But some technical points are worth mentioning:

(i) It may be convenient to use, explicitly, different kinds of higher type functionals in the formalization of analysis; cf. the work of Howard, cited [4], p. 141, 2.6244.

(ii) Besides the difference between continuity properties of $\forall\alpha\exists x$ and $\forall\alpha\exists\beta$, there is also a difference between so-called *weak* and *strong* continuity of $\forall\alpha\exists x$, i.e.,

$$\forall\alpha\exists x A(\alpha, x) \supset \forall\alpha\exists x\exists y\forall\beta[\alpha \mathrel{\underset{y}{=}} \beta \supset A(\beta, x)]$$

and

$$\forall\alpha\exists x A(\alpha, x) \supset \exists\gamma[C_0(\gamma) \ \& \ (\forall\alpha)A(\alpha, \Gamma\gamma)],$$

([4], 2.511, p. 135) where $C_0(\gamma)$ expresses that γ is a neighbourhood function and Γ is the functional determined by γ.

(iii) In the absence of the continuity axiom for $\forall\alpha\exists\beta$ the crucial contextual *elimination of free choice sequences* ([4], p. 140, 2.622) needs re-examination for the wider context of intensional operations. The difference between Myhill's analysis in his section 6 and the analysis in 2(c) of my reply to Heyting may be important: but I do not understand it very well. (It *seems* that he considers not only free choice sequences of natural numbers, but also of constructive functions such as spreads. If this is really the essential difference, the *formal* general theory would not be affected as long as one only uses the familiar existential axioms for constructive functions; cf. [4], pp. 141–142).

2. *The theory of the thinking subject.* In view of my proof above of $\sim\forall\alpha[\sim\sim\exists x(\alpha x=0)\supset\exists x(\alpha x=0)]$ without appeal to $\vdash_n$, Myhill's results (and not section 3 of my paper) seem to me the first genuine results in the theory of $\vdash_n$. He has obtained clear, meaningful consequences of the axioms for $\vdash_n$, consequences which are certainly not formally derivable in previously formulated systems (because they are false for the intended interpretation of those axioms). It would be very interesting to decide whether this theory is a conservative extension of known systems in the following precise sense: If A is an assertion in the language of intuitionistic analysis in which all quantifiers are interpreted extensionally and A is formally proved by use of the axioms for $\vdash_n$, can it also be proved without these axioms? (The requirement on the quantifiers is of course essential, by Myhill's section 1.)

3. *The proof of* $\sim\forall\alpha[\sim\sim\exists x(\alpha x=0)\supset\exists x(\alpha x=0)]$. The argument in the text gives a correct deduction of

$$\forall\beta[\sim\sim\exists x(\beta x=0)\supset\exists x(\beta x=0)]\supset\forall\beta[\exists x(\beta x\neq 0)\vee\forall x(\beta x=0)]$$

by essential use of the axioms for $\vdash_n$ (essential since the hypothesis is applied to an empirically defined sequence α). But it is a sheer 'fluke' that $\sim\forall\beta[\exists x(\beta x\neq 0)\vee\forall x(\beta x=0)]$ holds for the intensional interpretation of the logical constants here used: this is only shown by Myhill's section 6; i.e. when I appealed to the absurdity of $\forall\beta[\exists x(\beta x\neq 0)\vee\forall x(\beta x=0)]$ I had in mind the proof which makes use of the extensional interpretation of the quantifier. So, though I knew of course the distinction ([4], 2.33) I simply did not pay attention to it in this context. Thus even for the case of an intensional interpretation of the quantifiers, the proof given in the text is not complete; it is only completed by Myhill's additional considerations.

It would be amusing if somebody got up and told us that the contradiction between the axioms for the thinking subject and the familiar continuity axioms showed the weakness of our informal notions! Amusing in view of the almost universal resistance to considering the new notion at all (cf. Heyting's comment) and in view of the universal acceptance of the distinction (between extensional and intensional operations) which is illustrated by the contradiction. We'd have the situation discussed on pp. 143–145 all over again!

What the work really shows is that the analysis (or, if one prefers, the correction) of our intuitive notions concerning free choice sequences was essentially *unambiguous*, and, I think, it is this side of it that is so delightful.

I had already thought a bit about Myhill's comparison between these notions and that of set. I don't know if the next generation will be interested in free choice sequences. But if it is they (or, at least, free choice sequences of constructive objects) will be at least as transparent as the (iterative) notion of set. After all, the latter is notoriously problematic with respect to the number of iterations (axioms of infinity). What remains to be seen is whether the delicate, intensional aspects of the notion of free choice sequence will be as fruitful for intuitionistic analysis as the delicate axioms of infinity seem to be for classical analysis.

FOUNDATIONS OF MATHEMATICS – WHITHER NOW?

LÁSZLÓ KALMÁR

József Attila University, Szeged

1. Mathematics was one of the first sciences [1] to employ *deductive inference*. Originally mathematicians used deductive inference as it was to be used by other scientists, to display rational connections between empirical facts; these connections enabled them to survey those facts more easily, and meant that they did not have to check each of them separately in practice. Thus deductive inference was an excellent tool with which to develop mathematics.

At about the same time mathematicians invented, or rather borrowed from philosophy, the method of *abstraction*. Originally they used this method in two ways: (i) to enable them to discover general regularities among empirical facts, by abstracting from accessory conditions which do not affect the regularity in question; (ii) to enable them to construct idealized concepts, e.g. point or straight line, which are realized only approximately in the outside world, but which, owing to their logical simplicity, are more suitable for use in deductive inferences than more complicated concepts which reflect more accurately the real world [2].

---

[1] Most mathematicians, including some historians of mathematics, believe that actually deductive inference was *invented* by mathematicians. However, A. Szabó has pointed out the impact of Elean dialectical philosophy upon ancient Greek mathematics, by showing how some mathematical terms, especially those relating to the deductive method, derive from Elean dialectic. (See his 'Greek Dialectic and Euclid's Axiomatics', this volume, pp. 1–8, and his earlier papers cited there.) Philosophy thus seems to have invented deductive inference before mathematics – unless, of course, future historical research shows that the Eleans borrowed it from earlier mathematicians.

[2] The usefulness of such idealized concepts stems from the fact that our knowledge, on all levels, consists of relative truths which coincide only approximately with reality (as Sir Karl Popper would say, all our knowledge consists of guesses which are always to be improved). Such relative truths

In Platonic philosophy universal concepts, especially those formed by abstraction used in way (ii), above, were admitted as 'ideas'; this rescued them from the common sense objections against their unreality, and promoted the development of mathematics until mathematicians (and philosophers) could understand better the essence, and usefulness, of abstraction.

In this way mathematicians, rather than investigating empirical facts, began to investigate the properties of those abstract concepts which fell within the scope of mathematics, without enquiring into their empirical origins. And instead of beginning their deductions from empirical facts, tested previously by observation, mathematicians began to use as premises intuitively evident properties of their abstract mathematical concepts – nobody enquired why they seemed intuitively evident, or why we should rely on intuitively evident propositions.

By basing mathematics on deduction from explicitly displayed intuitively evident propositions, mathematicians invented the *axiomatic method*.

2. The invention of deductive inference, of abstraction, and of the axiomatic method, was very fruitful for the development of mathematics. But it tempted mathematicians to regard mathematics as a 'pure deductive science', and to forget that their axioms were originally abstracted from empirical facts, and that their rules of deductive inference are valid because they have been tested in the actual thinking practice of mankind.

Several circumstances combined to sustain the view that mathematics is a 'pure deductive science': (i) the rise of a philosophy of the *a priori*, according to which it was unnecessary to test in practice intuitively evident axioms and rules of inference; (ii) mathematics' becoming proverbial as an 'infallible' science, due to its early luck in finding its appropriate methods; (iii) mathematics' becoming, because of its precision, an ideal for other fields, be-

---

can contain concepts which correspond only approximately to reality; and as our guesses improve, we can construct from our idealized concepts new concepts which are 'less idealized', or which correspond more closely to reality (e.g. that of a geometrical figure as a set of points in space).

ginning perhaps with Spinoza's *Ethica more geometrico demonstrata*, down to the mathematical economics or mathematical linguistics of our own day; (iv) the success of mathematics itself, and the extension of its method of abstraction to more and more general regularities holding for empirical facts, which was reflected in more and more abstract mathematical theorems and algorithms.

3. Despite this, the infallibility of mathematics, and of its basis in deductive inference, had already been questioned in antiquity, the former, for example, by Zeno the Eleatic, and the latter, for example, by Zeno the Stoic. But Zeno's paradoxes did not refer to 'pure' mathematics, and so did not seriously shake the dogma that mathematics is a self-contained pure deductive science. And, moreover, the connections between logic and mathematics were not considered so essential in antiquity as they are nowadays (especially since Logicism came into being), and so the Sophist's criticism of logical deduction did not affect the dogma of the purely deductive character of mathematics.

Later the paradoxes of the infinitesimal calculus, especially those connected with calculation with divergent series, again called the infallibility of mathematics into question. Mathematicians, who could use 'Newton's and Leibniz's deep concepts' to get correct results, took these paradoxes very seriously and this grain of sand in the works irritated mathematicians into producing a real pearl – the Cauchy–Weierstrass Infinitesimal Calculus. In abolishing infinitesimals in favour of what were, in effect, set-theoretic concepts, mathematicians could once more claim that mathematics is an infallible deductive science, against those who had doubted it because of the paradoxes of the Calculus.

But the secretion of this pearl gave only a brief respite to our mathematical shellfish. Soon new sand-grains, the set-theoretic Antinomies, began to irritate them. And our mathematical shellfish, with their different degrees of sensitivity, produced several new pearls to get rid of the new irritation – these new pearls constitute the so-called Foundations of Mathematics.

4. One of these pearls, Type Theory, is one of the most ingeniously sophisticated systems the human mind has ever produced. The type-restrictions, Whitehead's and Russell's commandments engraved in their Two Tables, were designed to make mathematics safe deductive consequences of trivial logical axioms, by forbidding incestuous relationships between, for example, a relation and the objects for which it holds. But the non-logical axioms of infinity and reducibility turned out to be indispensable – and with this Type Theory became nothing more than one possible system in which to axiomatise a part of mathematics, and thus, like any other formal system, in need of a consistency proof to ensure that no new antinomies would emerge. Moreover, the new formal systems of Axiomatic Set Theory showed that a formal system which obeyed the type-restrictions contained only a modest (though important) part of mathematics. The type-restrictions give us a pretty uncomfortable formal system – they certainly violate Bernays' Principle [1] that we should make our logic as comfortable as possible.

Quine's ingenious idea of stratification [2] gave new hope to Type theorists; but the loosening of the type-restrictions involved in it was not radical enough to fulfil these hopes.

5. A second mathematical pearl, Intuitionism, sacrificed large parts of mathematics in exchange for the soothing assurance that what remained was justified by our 'primordial intuition' (*Urintuition*). But intuition is subjective, and not intersubjective enough to prevent intuitionists from differing about what their 'primordial intuitions' should enshrine as the basis of mathematics. So the one true intuitionist Church has disintegrated into several nonconformist sects, each with their own heresies. Members of most of these sects are condemned to suffer eternal damnation in the Hell of the Denumerable, because their 'primordial intuition' does not justify a belief in the Free Choice Sequence which alone could save them from it.

---

[1] Formulated in his paper 'What do some recent results in set theory suggest?', this volume, especially p. 112.

[2] See W. V. Quine, 'New Foundations for Mathematical Logic', *The American Mathematical Monthly*, **44** (1937), pp. 70–80.

6. Another pearl from the Foundations of Mathematics, Hilbert's Proof Theory, seemed at first to provide a powerful method of assuring mathematicians that no new kind of logical antinomy would emerge. And for this, they had only temporarily to sacrifice the meaning of mathematical theorems, treat them as mere strings of primitive symbols, and avoid non-finitistic inferences in Metamathematics. After a brief sojourn in the Purgatory of consistency proofs, mathematicians would be free once more to attribute meaning to their formulae, and to use any rule of inference of a system which had been proved consistent.

Gentzen's proof of the consistency of the Arithmetic of the natural numbers is the most outstanding positive result of Proof Theory. But since Gentzen, Proof Theory has turned out to be more suitable for achieving *negative* results about the categoricity and monomorphism of formal systems (except for those who are contented with positive 'second order' results). The most important of these negative results are Löwenheim's and Skolem's theorems, Gödel's incompleteness theorems, the existence of non-standard models, and Gödel's and Cohen's results on the independence of Cantor's Continuum Hypothesis in the usual axiomatic systems of Set Theory. These negative results showed that we have to give up the classical idea that the primitive ideas of a branch of mathematics can be implicitly defined by a system of axioms; instead, we now think that an axiom system defines the common properties of all models of it, standard and non-standard, a point of view adopted long ago in Algebra. Soon we will speak of 'an arithmetic of natural numbers' or 'a set theory', just as we now speak of a group or a ring.

A formal system, it now appears, can serve merely to display some set of formulae as a recursively enumerable set, by showing it to be identical with the set of theorems of the formal system in question, rather than to facilitate a consistency proof for a branch of mathematics. It seems unlikely that, even in the best case, we shall extend the field of demonstrably consistent mathematics beyond classical analysis – unless computers, using induction (or some other) principles too sophisticated for the human mind, can help us in constructing consistency proofs.

7. At any rate, Foundations of Mathematics, after flourishing vigorously in the first half of our century, when everyone was optimistic about achieving quick results in several directions, now seems to be in a period of pessimistic stagnation, of withering, despite some interesting results achieved recently. Hence it is appropriate to ask, using a pun [1]: Foundations of Mathematics – whither now?

To prophesy is always risky. So I do not want to give any definite answer to our question. No doubt, people will continue to walk the present avenues of research, and new interesting results may be achieved. But I do not think that in this way we can turn a withering plant into a flourishing one. In order to find new avenues of research, we have to face some facts. Firstly, research in the Foundations of Mathematics has hitherto been pursued with the supposition that mathematics is a pure deductive science, and with the hope that we shall show it to be a *firmly founded* pure deductive science. Secondly, this hope has never, in fact, been realised [2]; and, as I pointed out above, the axioms of any interesting branch of mathematics were originally abstracted more or less directly from empirical facts, and the rules of inference used in it have originally manifested their universal validity in our actual thinking practice. Thirdly, the consistency of most of our formal systems is an empirical fact; even where it has been proved, the acceptability of the metamathematical methods used in the proof (e.g. transfinite induction up to some constructive ordinal) is again an empirical fact. Even if we continue to try to reduce mathematics to intuitively evident logical or other principles, and do not admit that what we regard as intuitively evident is ultimately a product of practical testing in experience, we must admit that it is an empirical fact that we can rely on evident principles. The validity or invalidity of Church's thesis, which is an important tool for proving some

[1] Used earlier by Kreisel, 'Informal rigour and completeness proofs', this volume, pp. 138–171.

[2] For related ideas on the hope, and its non-realization, see I. Lakatos, 'Infinite Regress and the Foundations of Mathematics', *Aristotelian Society Proceedings, Supplementary Volume* 36 (1962) pp. 155–184, and 'Proofs and Refutations I–IV', *British Journal for the Philosophy of Science*, **14** (1963–4).

negative statements in the Foundations of Mathematics, is once more an empirical question [1].

Why do we not confess that mathematics, like other sciences, is ultimately based upon, and has to be tested in, practice? Many respectable sciences have excellent reputations, without claiming that they are 'pure deductive sciences'. To declare mathematics a science having an empirical ground would not exclude the use of deductive methods, for many other empirical sciences use them successfully! True, we would then have to include in Mathematics inductive methods – but why exclude them? Is it because we cannot be sure that a *general* statement, proved by *induction*, holds true in all cases? Well, deduction is no more safe in this respect: we cannot be sure that an *existential* statement, proved by *deduction*, holds true. There are pure existential proofs, which have provoked intuitionist criticisms, and there are formal systems which are not, or have not been proved to be, ω-consistent. Whether or not we admit inductive methods to mathematics, or confine ourselves to deduction (except for establishing the axioms or rules of inference), our mathematical theorems will be (at least partly) relative truths, perhaps requiring modification in the future. They can, however, be good approximations to absolute truths (i.e. to what is the case in reality) just as some physical laws are, or are guessed to be on the basis of practical evidence.

To regard mathematics as a science in need of an empirical foundation will give rise to new problems in the empirical Foundations of Mathematics. I guess that research on these problems will be one of the main avenues of research in the future, if not the main avenue. I cannot now specify all these future avenues of research; hence the question in the title of my paper. But I can indicate some of the problems which we will have to face.

---

[1] In my paper 'An Argument against the Plausibility of Church's Thesis', in *Constructivity in Mathematics*, edited by A. Heyting, Amsterdam, 1959, pp. 72–80, I pointed out that plausible arguments can be given *against* Church's thesis as well as for it. But I have no objection against Church's thesis if it is taken as an *empirical* one, confirmed several times in practice, but, like any other empirical thesis, to be abandoned if a counter-example is found in the future.

For example, what are the 'observables' in empirical mathematics, i.e. what kind of mathematical proposition can be tested directly in practice? Obviously, pure numerical propositions are of this kind; and in some sense so are, for example, propositions formed from them by replacing numerals by variables bound by universal quantifiers. Another problem is: how can we indirectly test those propositions which cannot be directly tested? Or, how can we classify mathematical propositions according to the degree to which they have been tested? Or, to what use can we put computers in empirical (or experimental) mathematics, apart from the accumulation of vast amounts of empirical evidence in support of some mathematical propositions?

I am convinced that once civic rights are granted to such problems, and they are investigated intensively, these regions of research will flourish.

DISCUSSION

A. Heyting: *Weyl on experimental testing of mathematics.*

The idea that mathematics should be subject to experimental verification has also been brought forward by Hermann Weyl [1]. In my opinion Weyl's conception goes deeper than Kalmár's. Weyl starts from the remark that an experiment never falsifies a particular physical assertion, but always a whole theory. He now proposes to integrate the formal mathematical apparatus into the physical theory. If then an experiment gives a result which contradicts the prediction from the theory, the theory must be modified, but we do not know where and how. It is conceivable that in order to preserve simplicity and coherence, we shall be led to modify the mathematical part of the theory. In this sense, as Weyl puts it, we might one day be induced to reject Hilbert's ε-axioms. By these means Weyl rescued formalistic methods from the blame of being nothing but a senseless play with signs. He felt that mathematics, if taken apart from physics, should be studied along intuitionistic lines.

S. C. Kleene: *Empirical mathematics?*

Professor Kalmár has been careful to say that he cannot now specify all the future avenues of research on the empirical foundations of mathematics.

Nevertheless, it seems appropriate to ask: How different is what Kalmár has in mind from what we're doing now?

For example, we are all familiar with evidence for Fermat's 'last theorem' based on computation, and theoretical work, that have confirmed it way, way up. Are we now to take this proposition

[1] *Philosophie der Mathematik und Naturwissenschaft*, Teil I; Handbuch der Philosophie IIA, München und Berlin 1926. Revised and augmented English edition: *Philosophy of Mathematics and Natural Science*, Princeton 1949.

as empirically established, to a degree; and add it to the stock of results we take as available for drawing conclusions? Is not this just what is being done now, though not actually I think with Fermat's 'last theorem'? But there certainly are people who have proved that, if Riemann's hypothesis holds then such and such holds, or if the axiom of choice holds then such and such holds, and these are interesting results. But while the state of the evidence is what it is now, it is the custom to put such suppositions explicitly into the hypothesis. Some other suppositions, such as the basic validity of the laws of logic – classical logic if you please – we feel more confident about, and these suppositions are made so often – on the basis of empirical evidence if you wish – that they aren't usually stated. But anybody who is being quite explicit about the theory he is developing will generally indicate at the beginning what his starting point is; or there will be indications in his writing from which it can be recognized. Mathematical developments proceeding from different starting points, such as from Riemann's hypothesis or from classical logic, will be evaluated relative to whatever kind of empirical evidence we feel we have for their starting points, just as is the case in physics.

PAUL BERNAYS: *Mathematics and mental experience.*

Professor Kalmár's thesis that mathematics is an empirical science sounds highly paradoxical. But the air of paradox disappears if we distinguish between a narrower and a wider sense of the word 'empirical'. Natural science is empirical in the sense that we come to our statements and test them by *sense-experience.* Now mathematics may have originated, psychologically speaking, from sense-experience, but we do not rely on it in the way we do in natural sciences. Indeed this is not possible as mathematical statements are not – at least not directly – about the external world.

However there are still analogies, especially stressed by George Pólya, between mathematics and empirical sciences as to the methods of heuristic trying and of testing. Moreover generally mathematical investigation has in so far an empirical character, as we do not know the facts from the beginning, but have to learn them, and that surprising discoveries occur in mathematics. But all these experiences come about not by external experimentation but rather by computing and inferring. Thus we may speak here of *mental experience*. Extending the term 'empirical' to mental experience, we might agree with Kalmár's thesis that mathematics is an empirical science.

Yet this way of speaking seems unsatisfactory in so far as it suggests a distinction of empirical and non-empirical sciences, which, by the widened use of the term 'empirical', becomes doubtful. We rather are induced to adopt Ferdinand Gonseth's *principle of duality* according to which in all scientific research a union of empirical and of rational elements is to be found.

Y. BAR-HILLEL: *Is mathematical empiricism still alive?*

I am afraid that I can agree neither with Professor Kalmár's 'highly paradoxical' view (as Professor Bernays puts it in his comments) nor with the remarkable degree of goodwill shown towards this view in the preceding comments.

But since Kalmár's view has in essence been propounded many times before in the history of philosophy of mathematics and has as often been refuted before, and since I have no new brilliant ideas in this connection (nor the feeling that such are needed for this purpose), let me dwell on a few minor points:

1. When Kalmár calls Type Theory 'one of the most ingeniously sophisticated systems the human mind has ever produced', I believe a distinction has to be drawn. The so-called Simple Theory of Logical Types seems to me to be a very simple and almost simple-minded matter, though not quite as trivial as Russell himself – perhaps only as an extreme case of the British tendency towards understatements – often made it look. The Ramified Theory is, of course, indeed sophisticated but still, in my estimate, a far cry from 'one of the most ingeniously sophisticated systems the human mind has ever produced'.

2. With all the admiration for computers, I regard Kalmár's remark 'unless computers, using induction (or some other) principles too sophisticated for the human mind, can help us in constructing consistency proofs' as close to frivolous, though it can be saved from this evaluation by being treated as just unintelligible.

3. Isn't it really time to stop using that phrase 'proof by induction' when one is *not* referring to proof by complete or transfinite induction? Kalmár falls immediately into the trap he has set himself by using this phrase. Nobody objects to the use of 'induction' (and other 'heuristic' techniques) for coming up with sentences that are worthwhile proving and even for getting some ideas about how to conduct the proof, but nobody – including Kalmár, I am quite sure – seriously wants to 'include in Mathematics inductive methods'. As a matter of fact, I have no idea exactly what this phrase is supposed to mean in this context.

4. No 'new' problems in the Foundations of Mathematics will be posed by an 'empirical' approach. Mathematicians, and other people, have been worrying for centuries 'what kind of mathematical proposition can be tested directly in practice' and others, like myself, will continue to believe that no mathematical proposition can be tested *directly* in practice.

5. I have some doubts whether Professor Popper would want to extend his conception of scientific knowledge – as consisting of guesses that can constantly be improved – to mathematics and whether he would think that concepts like 'degree of testability' were applicable to mathematics?

6. Weyl's conception of mathematics, as explained by Professor Heyting, seems to me to be as correct as can be put in five sentences. The only thing I do not understand is why mathematics should be interpreted (I take this to be the intended term, rather than 'studied') apart from physics (i.e., presumably, empirical science) at all – with the consequence that this interpretation has then to be done along intuitionistic lines. In my conception, mathematics receives its interpretation just by being part of some interpreted empirical theory.

7. I continue to be unhappy about the phrase 'mental experience' which Professor Bernays and others are fond of using

in connection with mathematics. I am similarly unhappy with the phrase 'testing mathematical statements'. What happens to a mathematical theorem which nobody has been testing in his mental experience for, say, three years? And what is the status of those many theorems that Professor Bernays is going to prove next year, for the first time in human history?

I. LAKATOS: *A renaissance of empiricism in the recent philosophy of mathematics?*

Professor Kalmár proposes that mathematics is an empirical science: its *origins* are empirical, its *method* is (and should be) similar to scientific method, its *justification* cannot exceed the justification of scientific theories.

Professor Bar-Hillel asks incredulously: 'Is mathematical empiricism still alive?' and seems to be perplexed by the 'remarkable degree of goodwill' shown to this so many times refuted view in Heyting's, Kleene's and Bernays' preceding comments.

True, the radical contraposition in the logic of justification advocated by logical empiricism between infallible *a priori* and tautologous mathematics on the one hand and fallible, *a posteriori* and contentful science on the other still dominates the philosophy textbooks. It may therefore come as a surprise to the historian of ideas when he finds statements by some of the best contemporary experts in foundational studies that may sound like heralding a renaissance of Mill's radical assimilation of mathematics to science. Kalmár's position is far from being exceptional.

Russell was probably the first who thought that the only evidence for mathematics and logic was 'inductive'. He, who in 1901 still claimed that the 'edifice of mathematical truths stands unshakable and inexpugnable to all the weapons of doubting cynicism', in 1924 thinks that logic (and mathematics) is exactly like Maxwell's equations of electrodynamics: both 'are believed because of the observed truth of certain of their logical consequences' [1].

---

[1] *Principia Mathematica*, Vol. I, p. 59 (Second Edition) and 'Logical Atomism', in J. H. Muirhead (ed.): *Contemporary British Philosophy*: *Personal*

Fraenkel, in 1927, claims that 'the intuitive or logical self-evidence of the principles chosen as axioms [of set theory] naturally plays a certain but not decisive role; some axioms receive their full weight rather from the self-evidence of the consequences which could not be derived without them'. And he compares the situation in set theory in 1927 with the situation in the infinitesimal calculus in the 18th century recalling d'Alembert: '*Allez en avant, et la foi vous viendra*', pointing out that it is a false belief that in mathematics such situations do not occur [1].

Carnap, who at the 1930 conference in Königsberg still thought that 'any uncertainty in the foundations of the "most certain of all the sciences" is extremely disconcerting' [2], in 1958 thinks that there is – if only a distant – analogy between physics and mathematics: 'the impossibility of absolute certainty' [3].

One could easily fill quite a few pages with similar quotations from Weyl, von Neumann, Bernays, Church, Gödel, Quine, Rosser, Curry, Mostowski and others in order to show that Kalmár's position – mathematical empiricism and inductivism (not only as regards the *origin* or *method* but as regards the *justification* of mathematics) – is more alive and widespread than Bar-Hillel seems to think. But what is then the background and what is the rationale of this new empiricist-inductivist *mood*? I think one can give it in terms of an epistemological demarcation between 'quasi-Euclidean' and 'quasi-empirical' theories.

The classical epistemological ideal of a theory – be it scientific or mathematical – was modelled for two thousand years on Euclidean geometry: a deductive system with an indubitable truth-injection

---

Statements, 1924, p. 362. He obviously hesitated whether one can put up with this state of affairs (and work out some sort of inductive logic for the *Principia*) or it is intolerable and one has to go on with the search for self-evident axioms. For more details about Russell's turn *cf.* my 'Infinite Regress and the Foundations of Mathematics', *Aristotelian Society Proceedings, Supplementary Volume* 36 (1962).

1 *Zehn Vorlesungen über die Grundlegung der Mengenlehre*, 1927, p. 61.

2 Die logizistische Grundlegung der Mathematik, *Erkenntnis*, 2, 1931, p. 91. English translation in Benacerraf–Putnam: *Philosophy of Mathematics*, 1964, p. 31.

3 Beobachtungssprache und theoretische Sprache, *Dialectica*, 12 (1958).

at the top (the conjunction of axioms) – so that truth, flowing down from the top through the safe truth-preserving channels of valid inferences, inundates the whole system.

It was a major shock for over-optimistic rationalism that science – in spite of immense efforts – could not be organised in such *Euclidean theories* [1]. Scientific theories turned out to be organised in deductive systems where the *crucial* truth-value injection was not truth at the top but falsity *at the bottom* – at a special set of theorems ('basic statements'). Thus the logical channels of such *quasi-empirical theories* are not for downward transmission of truth, but for upward retransmission of falsity – from the basic statements to the axioms.

Perhaps one can characterise Euclidean theories best as limiting cases of quasi-empirical theories: a quasi-empirical system is Euclidean if it is the logical closure of the accepted basic statements and (strictly) quasi-empirical if it is not. Any quasi-empirical system has then a Euclidean kernel.

A Euclidean theory may be claimed to be true; a quasi-empirical theory – at its best – to be well-corroborated, but ultimately conjectural. Also, in a Euclidean theory the axioms *prove*, as it were, the theorems; in a quasi-empirical theory the axioms *explain* the theorems.

Whether a deductive system is Euclidean or quasi-empirical is then decided by the direction of truthvalue flow in the system. This demarcation has nothing to do with the particular form of the axioms and of basic statements or with the particular conventions that regulate the truthvalue injection. For instance a quasi-empirical theory may be empirical or non-empirical: it is empirical only if its basic statements are singular spatio-temporal statements whose truthvalue is decided by the time-honoured but unwritten conventions of experimental science. But then we may, still more generally, abstract from *what* flows in the logical channels, certain or fallible, objective or conventional truth and falsehood, probability and improbability, moral desirability and undesirability,

[1] For an exposition of the story see my 'Infinite Regress and the Foundations of Mathematics', *Aristotelian Society Proceedings, Supplementary Volume* 36 (1962).

etc.; it is the *how* of the flow that is decisive. This will yield us quasi-Euclidean and (generalised) quasi-empirical theories.

The methodology of a science strongly depends on whether it aims at a quasi-Euclidean or at a quasi-empirical ideal. The basic rule of the former is to search for self-evident or at least plausible axioms – Euclidean (or quasi-Euclidean) methodology is puritanic, antispeculative. The basic rule of the latter is to search for bold, imaginative hypotheses with high explanatory and heuristic power; indeed, it advocates an uninhibitedly speculative proliferation of alternative hypotheses to be pruned by severe criticism [1].

Now my thesis is this: the present empiricist mood, so strongly represented by Professor Kalmár, originates in the recent recognition that mathematics is not Euclidean or quasi-Euclidean – as had been expected in the first heroic period of foundational studies – but quasi-empirical, and in the increasing attention drawn to the part of mathematics that falls *outside* its Euclidean kernel.

But if mathematics and science are both quasi-empirical, the crucial difference between them, if any, must be in the nature of their 'basic statements', or 'potential falsifiers'. *The 'nature' of a quasi-empirical theory is decided by the nature of the truthvalue injections into its potential falsifiers.* I do not think anybody would claim that mathematics is *simply and directly* empirical in the sense that its potential falsifiers would be singular spatio-temporal statements. *But then what is the nature of mathematics? Or, what is the nature of the potential falsifiers of mathematical theories?* [2] This quasi-Popperian formulation of the age-old question may very well shed some new light on a few questions in the philosophy of mathematics – including, for instance, the problem of the refutability of Church's thesis which was raised so sharply, by no accident, by Professor Kalmár.

---

[1] The elaboration of empirical methodology – which of course is the paradigm of quasi-empirical methodology – is due to Karl R. Popper.

[2] For an elaboration of this and related problems, *cf.* my 'Empiricism in the Contemporary Philosophy of Mathematics', *British Journal for the Philosophy of Science*, **18**, 1967.

L. KALMÁR: *Reply.*

I am much honoured by Heyting's comparison of my conception of mathematics with H. Weyl's. I admit that, from the point of view of the philosophy of physics, Weyl's conception goes deeper than mine. However, our present problem is the foundations of mathematics. From this point of view, I cannot consider Weyl's conception deeper than mine; as a matter of fact, I cannot consider it even a consistent one. Weyl, as a mathematician, was ready, as any intuitionist, to sacrifice large parts of mathematics in order to justify the rest – in terms of his extremely high standards of rigour. On the other hand, Weyl, as a physicist, was not ready to sacrifice them, for they are useful tools in physics. He does not explain how some propositions can be useful tools in one science if they are definitely false in another, the very science to which they belong – as indeed they are, according to Weyl's mathematical views. I hope that future research work in the empirical foundations of mathematics will show in which sense propositions rejected by intuitionists can be tested in practice, and in which sense such a testing justifies their application in physics or other sciences.

Professor Kleene asks how different is what I have in mind from what we are doing now? Of course, schools which want to change present mathematical practice by forbidding something, like Type Theory or Intuitionism, can easily reply to such a question by displaying examples of what has been done so far but which will be forbidden from now on. A school which, instead of requiring sacrifice, wants to give a firmer foundation to the whole than had existed before, can only reply: we shall – at least in pure mathematics – do the same as we are doing now, but with a better conscience. So I am convinced that we can continue our research-work in mathematics with a better conscience if we confess that different parts of its fundamentals are based on empirical evidence, than if we deceive ourselves in saying that they are finally well-founded.

I agree with Professor Bernays that mental experiences play an important role in mathematics. In my paper, I pointed out that the rules of inference which we apply in most formal systems actually used in mathematics have originally manifested their universal validity in our actual thinking practice. I admit also, that even in cases where we can test a mathematical proposition (e.g. a numerical one) by means of a 'sense experiment' done with concrete objects, we often replace the latter by a 'mental experiment' done with the abstract reflections of these objects in our mind. However, I want to stress that this replacement of a sense experiment by a mental one is also based on the fact, tested several times in practice, that what has been tested by such a mental experiment can be tested by the corresponding sense experiment as well.

As to Gonseth's principle of duality, I stressed in my paper that not only mathematics, but other sciences as well, need deductive inference, besides empirical facts, to display rational connections between empirical facts in order to be able to survey them more easily and to manage without checking each of them separately in practice.

I do not regard Professor Bar-Hillel's disagreement with what he calls 'the remarkable degree of goodwill' shown towards my view by the authors of the preceding comments as a sign of a lack of good will towards those who hold philosophical views opposed to his own. On the contrary, I know that Bar-Hillel is personally of a friendly disposition; hence, I must attribute the difference in style between his comments and those of Heyting, Kleene and Bernays to the fact that the latter are mathematicians whereas Bar-Hillel is a philosopher. In saying this, I do not want to imply that when a philosopher finds somebody whose views are opposed to his own views and moreover, who refers to scientific arguments to support those opposite views, he necessarily must ask whether such opposite views are still alive. As a matter of fact, there are philosophers who respect scientific facts to such a high degree that they are willing to change their views if they feel that some new facts of science, or failure of the hopes which served as a basis of their old views, force them to do so. Obviously, Bar-Hillel is not willing to change his view that mathematics cannot be founded on experience even in the light of the facts I mentioned in my paper, nor does he wish to consider seriously the views of several mathematicians and philosophers, quoted by Lakatos, which are closer to my views than to his. Hence, he has no other choice than to be angry that somebody draws the conclusions which I did from these facts.

As to the distinction between the Ramified Type Theory and the Simple Theory, of course, it was the first one to which I referred in my paper, for actually, it was the first pearl produced by Russell and Whitehead when they were irritated by the set-theoretical antinomies. The Simple Type Theory was a later product, due to the feeling that the Ramified Theory needed simplification. Its creation shows that even Type theorists considered the Ramified Theory rather sophisticated. If Bar-Hillel does not share my admiration of the Ramified Theory as one of the most ingeniously sophisticated systems the human mind has ever produced, I would ask him to give some argument.

In any case, Simple Type Theory did not succeed either in showing that mathematics can be entirely reduced to logic, even by sacrificing intensions, which seem not only philosophically, but sometimes even mathematically important. (E.g. we must distinguish two arithmetical functions defined by different recursions because we have in general no algorithm to decide whether two recursions define extensionally identical functions or not.)

As to the idea of using computers (in the future) for constructing proofs based on principles too sophisticated for the human mind, I do not understand why Bar-Hillel calls it 'close to frivolous' or 'unintelligible'. Does he consider 'close to frivolous' or 'unintelligible' the idea of using a microscope to observe objects too small for human eyes? Why exclude *a priori* the possibility of 'lengthening' the human mind by means of artefacts if we can 'lengthen' human senses so?

I do not regard (non-complete and non-transfinite) induction as a mere heuristic technique for coming up with sentences that are worthwhile proving, but as one of the means of gaining relatively true sentences which we may trust until they are disproved. I am convinced that our deductive methods (e.g. proof in classical analysis) are just another (if you want, in some respects more effective) means to the same end and not for gaining 'finally founded absolutely true sentences' – at least as long as no one proves the consistency of classical analysis and convinces me that the metamathematical methods used in his consistency proof are 'finally acceptable'.

Obviously, Bar-Hillel means by a proposition which can be tested directly in practice a proposition for which one such test makes any further test unnecessary. For otherwise, why does he not admit that if we form the union of two disjoint sets each of two concrete objects and count the number of the objects in the union, then we use a direct way of testing in practice the proposition $2 \cdot 2 = 4$? However, I do not believe that even a billion tests each confirming a proposition in practice, make further tests unnecessary.

I cannot answer for Professor Popper Bar-Hillel's question about whether he would want to extend his conception of scientific knowledge to mathematics. However, I am sure that he does not consider mathematics a shelter for those who want to live among indubitable truths.

In his remark on Heyting's comments, Bar Hillel agrees in interpreting mathematics as part of some interpreted empirical theory. Does this mean that he agrees to test interpreted mathematical propositions in practice? If so, why does he not allow us to construe a mathematical proposition as a product of abstraction from all of the corresponding interpreted propositions, and to test it in practice by testing the corresponding interpreted propositions one after the other?

My reply to his last question is the following. The theorems that Bernays is going to prove next year refer, presumably rather indirectly, to some properties of the real world, with which they are connected by a complicated chain of abstractions. These properties of the real world held, of course, before Professor Bernays has proved the theorems in question, and indeed, before any human being lived on earth.

I am indebted to Lakatos for showing by many quotations that my position, criticized by Bar-Hillel, is far from being exceptional. He explains the position by embedding it in his very interesting concept of quasi-Euclidean and quasi-empirical theories. I have only one doubt concerning this concept. He regards the logical channels for truth-value flow inside a theory as given in advance, whereas I consider the questions of whether these channels work at all, whether they admit downward flow of truth or upward flow of falsity, as questions to be decided by testing the truth-value flow in practice.

As to his last remark, I consider any proposition stating the absolute unsolvability of some problem (with parameter) proved on the basis of Church's thesis, as a potential falsifier of all theories which are based on this thesis, of course without being sure or even without suggesting that any of them will be falsified, by solving the problem in question using some acceptable (though of course, not recursive) method, some time in the future. A very interesting question is whether there are potential falsifiers of a related kind of theories based on some hypothesis related to, but not equivalent with, Church's thesis. In any case, the above-mentioned ones are good examples of potential falsifiers of mathematical – or rather metamathematical – theories, which are far from being singular spatio-temporal statements.

LOGIC AND HEURISTIC IN MATHEMATICS CURRICULUM REFORM [1]

JACK A. EASLEY, Jr.

University of Illinois, Urbana

1. *The 'Math Wars'*

The past decade has seen a revolution in American education which promises to lift it out of the doldrums of intellectual mediocrity in which it has been becalmed – perhaps fundamentally because of the very rapid expansion of opportunity for schooling. This is a revolution in methods of curriculum reform that is now going forward at an unprecedented rate. (See Goodlad [13] for an excellent review of this movement.) It was sparked by the vigorous participation in the design and preparation of instructional materials by numerous scientists and mathematicians of a considerably higher level of ability than have previously been engaged in curriculum work. Despite such unprecedented commitment of talent, the movement has encountered some sharp criticism – a fact which is perhaps not especially surprising. But it is puzzling that the criticism has been much sharper in mathematics than in the other sciences. The new mathematics programs, as they are viewed by certain members of the mathematics profession, are said to be completely wrongheaded. In spite of its sharpness, however, this criticism has contributed little to understanding the problems of mathematics teaching or what the next steps should be. The reformers and their critics give the appearance of being committed to diametrically opposed objectives and of being bent on forming completely independent schools of pedagogical practice.

---

[1] I wish to thank Maurice Tatsuoka, Robert Davis, Max Beberman, and Herbert Vaughan for their helpful suggestions but absolve them of all responsibility for the views expressed in this paper. I am most grateful to Nancy Zukas and Rose Vanerka for patiently typing numerous drafts.

One conjecture as to why mathematics is more controversial than other areas of the curriculum is that mathematicians are less well agreed than physical and biological scientists about the nature of their discipline. Whatever truth this hypothesis may have psychologically, I think it reflects a mistaken notion, as I shall later point out, of the relationship between philosophy of science and educational problems.

As an educationist, I am concerned to try to turn criticism of such a *divisive* nature into somewhat more *decisive* criticism. Continued battling out of the 'Math Wars' (cf. DeMott [9]) does not serve this end. It is necessary to examine the underlying issues of the controversy to try to discover whether they are subject to empirical investigation, philosophical investigation, or had best be met by practical compromises. I shall focus my attention on those issues which have to do with the role of logic, physical intuition, and heuristic procedures in school mathematics instruction, because these issues appear to lie close to the heart of the debate.

2. *The 'New Math'*

A few words are necessary about the groups of mathematicians, professional educators, and teachers who are engaged in this controversial curriculum reform movement in America. The first of these course-content improvement projects (to use the terminology of the National Science Foundation which now supports many of them) was organized at the University of Illinois by Max Beberman in 1951. During its first decade, this project produced a four-year series of mathematics texts for high schools. These texts were characterized by the informal introductions given to new topics but also by the precise language used in the formulation of theory. They also are designed to lead students to discover concepts and principles themselves, and make use of logical principles and set-theoretical concepts in clarifying them, once discovered. This project is now engaged in developing a remedial pre-algebra and intuitive geometry course for the junior high school, and a vector approach to geometry at the senior high school level.

A second major group, which has certainly been the most influential in terms of adoptions of its textbooks, was begun at Yale University in 1958 under the direction of Professor Begle, and is now operating at Stanford University under the same director. This group has produced texts for all elementary and secondary grades. These texts cover the standard topics from a modern point of view and introduce a number of new topics including operations with sets at some elementary grade levels, probability in grade 7, and matrix algebra in grade 12. (See Wooten [40] for a history of this project.)

Professor Suppes' project at Stanford University pioneered in building arithmetic from operations on sets and deserves special mention as the only project which has introduced constructive geometry and formal logic into the elementary grades. The University of Illinois Arithmetic Project directed by David Page, and the Madison Project of Syracuse University and Webster College, Missouri, directed by Robert Davis, have pioneered in introducing algebraic notation and concepts into the elementary grades. Other influential projects have arisen at the University of Maryland, at Ball State Teachers College in Indiana, at the University of Minnesota, at Boston College and in a number of other places.

These new mathematics projects have been influential in changing the American school curriculum in a great many small cities and suburban areas, but have had much less effect to date in areas of very high or very low population density. One of the largest practical problems this movement has encountered is the problem of the education and re-education of teachers in the understanding and appropriate use of mathematical concepts that are new to them. Another problem of considerable importance, however, is that of evaluation of the effects of these new programs on the mathematical abilities and interests of children. Because I do not have time to review the history of the movement in greater detail, it should at least be said that there are interesting differences among the various course-content improvement projects which I am forced to overlook. However, it *is* fair to generalize by saying that all of these American projects have gone a considerable distance in introducing concepts and techniques into the curriculum which reflect developments in the foundations of mathematics, particularly concerning the concept of set, set-theoretic models, axiomatic method and proof. I must turn now to an examination of the major criticisms of the mathematics curriculum reform movement.

3. *The critics' views*

The critics of the new mathematics programs are represented in breadth by 65 mathematicians who attached their names to an article published three years ago which raised major objections to the new programs. It specifically complained that they failed to follow pedagogical principles such as, 'the introduction of new terms and concepts should be preceded by sufficient "concrete" preparation', and that intelligent youngsters should not be expected to learn arbitrary rules. This article proposed that 'the feeling for rigor [in proofs] can be much better learned from examples wherein the proof settles genuine difficulties than from hair splitting or endless harping on trivialities.' (Ahlfors *et al.* [1]) So much for the general flavor of the criticism!

The views of these critics are represented in considerably greater depth by three of the 65 mathematicians whose articles and books express a point of view regarding the teaching of mathematics consistent with that expressed in the above-mentioned article. These three are Professors Morris Kline, Alexander Wittenberg, and George Pólya, each of whom is wellknown for his contributions in interpreting mathematics as a creative enterprise which interacts strongly with the sciences, as well as with other aspects of human culture (Kline [17, 18, 19], Wittenberg [36, 38], Wittenberg and Sœur Sainte Jeanne-de-France [39], Pólya [27, 28, 30, 31, 32]). I shall identify three issues which their criticism raises.

4. *The issue of formality – the role of proofs and logic*

First of all, the critics object to what they regard as the excessively formal treatment given most topics in the new mathematics programs. For example, in a recent collection of opinions on the subject (Moise *et al.* [22]), Professor Kline writes, 'Mathematics in these [modern] curricula is a collection of deductive structures. The correct approach in my opinion, is not deductive but constructive. We should start with simple and concrete situations and with heuristic arguments and only gradually arrive at a deductive organization.'

Regarding proof Kline says, 'The proper approach is to be as intuitive as possible, to adopt axioms which mean something to the student, and to prove only what the student thinks requires proof. The very function of proof is not to prove the obvious in order to meet some sophisticated and purely professional standards and goals, but to prove what is not obvious.' (Moise *et al.* [22]).

Regarding styles of proof, Pólya writes, '. . . I think that in teaching high school age youngsters we should emphasize intuitive insight more than, and long before, deductive reasoning. And when we present proofs, we should present them much closer to the idea of Descartes than to a certain idea of certain modern logicians. . . . Crowding the pages of a textbook with pointless proofs which lack motivation and rewarding goals may make the worst impression on the best students, who have some gift of intuition which could be most useful in engineering or science or mathematics.' (Pólya [32], pp. 128 ff.)

Summarizing the Cartesian conception of proof, Pólya writes, 'A mathematical deduction appears to Descartes as a chain of conclusions, a sequence of successive steps. What is needed for the validity of deduction is intuitive insight at each step which shows that the conclusion attained by that step evidently flows and necessarily follows from formerly acquired knowledge (acquired directly by intuition or indirectly by previous steps of deduction).' (Pólya [32], p. 127. See also Pólya [29] for an elaboration of this conception of proof.)

5. *The issue of a physical basis for mathematical theory*

The second major issue arises from the critics' claim that the new mathematics programs do not make sufficient use of physical systems from which students can intuitively discover mathematical relationships. Kline takes the philosophical position that 'The sense of mathematics is physical [and] without this sense we have nonsense.' (Moise *et al.* [22], p. 15). However, he interprets this position educationally, emphasizing the importance of what are conventionally called the physical applications of mathematics.

The word 'application', however, does an injustice to the critics, because it suggests to them the wrong pedagogical direction. Instead of first developing a body of mathematical ideas and only then introducing various physical applications of it, Wittenberg recommends that, 'the applications rather should arise naturally, simply because elementary mathematical ideas have a practical import and a practical meaning, and this the student should be made to grasp.' (Friedman [12], pp. 46 ff.) The emphasis given to physical and geometrical problems by the critics is clearly intended to provide intuitive guidance in problem solving. However, another source of guidance in problem solving is also urged by the critics.

6. *The issue of heuristic treatment of problem solving*

The third major criticism concerns the failure of the new programs to develop mathematics heuristically. We have quoted Kline's recommendation that instruction should proceed with heuristic arguments. The main thrust of this argument, however, comes from Pólya, whose detailed exposition of heuristic procedures in a variety of mathematical fields is well known [28, 30, 31, 32]. For example, in his book, *How to Solve It* [27], we find such helpful guides to solving problems as the following:

1. Is the condition sufficient to determine the unknown? Or is it insufficient? Or redundant? Or contradictory?
2. If you cannot solve the problem try to solve first some related problem. Could you imagine a more accessible related problem?
3. Did you use all the data? Did you use the whole condition?
4. Carrying out your plan of the solution *check each step*. Can you see clearly that the step is correct? Can you prove that it is correct?
5. Can you check the result? Can you check the argument? Can you derive the result differently? Can you see it at a glance?

Some of these guides, e.g. (2), have the linguistic form of rules, and all of them could be so rephrased. Therefore, the third criticism could be interpreted as indicating that students should be explicitly taught heuristic rules of the above sort. By heuristic rules, in general, are meant rules which may be expected to aid in solving problems but which are not guaranteed to succeed in doing so. In this respect they are to be distinguished, of course, from algorithms.

7. *The reformers' response to their critics*

The reformers have typically responded to their critics in rather general terms, but one can occasionally find a direct statement of their position on the first of the three issues. For example, in the introduction for teachers in their geometry unit, Beberman and Vaughan [4] respond as follows: 'It is sometimes asserted that to ask students to study proofs of "obvious" theorems is [necessarily] stultifying. On the contrary, for a student who already has some notion of proof and is in the process of enlarging this notion, such proofs serve as tests of the principles of logic which he is on the verge of accepting. Rather pragmatically, he argues that since the use of the principles enables him to prove some theorems which are intuitively correct, the principles are probably valid.'

The reformers have mainly responded to their critics by pointing to the success adequately prepared teachers have had in sustaining the interests of children in a subject which has the reputation of often being exceedingly dull and sometimes quite abstruse. They point to the achievement of practical goals such as learning to compute, but primarily to the vast improvement they have made in providing an intellectual basis for understanding mathematical ideas (Moise *et al.* [22], Begle [6]). Thus, Davis has written, 'a significant minority of schools are teaching a genuinely modified and improved version of "mathematics". Notations are clearer, definitions more wisely chosen, student participation is greater and is focused more toward honest essentials, the sequence of topics is improved, important and appropriate new topics have been added, relatively unimportant (or even wrong) material has been deleted, the pace is better adjusted to actual student potential, the quality of expository writing has been improved, relations to other subjects are more adequately treated, and – almost most important of all – a more creative and exciting flavor permeates the program.' (Moise *et al.* [22], p. 10.)

Little has been said in response to the second and third issues – the alleged failure of the reformers to use physical models and heuristic procedures. Absorbed in the task of carrying their programs to completion, the reformers have more often than not ignored their critics, thus leaving it to educators to make up their own minds about the controversy.

Although the opinion has been expressed that the dispute simply concerns a matter of taste (Baker, in Moise *et al.* [22]), this controversy cannot be allowed to rest on such a basis. Perhaps what is meant is that the two main positions in the controversy arise from fundamentally different philosophies of mathematics – perhaps embodying sharply different views of the foundations of mathematics, such as those Lakatos [20] has delineated – a formalist, dogmatic epistemology *vs.* an intuition-oriented, skeptical view of mathematical knowledge. Even so, the different recommendations for school curricula made by the reformers and their critics should be examined primarily in terms of the hypotheses about pedagogical matters that are implicit in them. These are hypotheses concerning, for example, what kinds of mathematical activities are of greatest interest to children and what kinds of mathematical activities are of greatest value in the development of problem-solving skill. Whatever one's view of the philosophy (or the foundations) of mathematics, such a view is ordinarily not directly applicable to the evaluation of a given piece of mathematics instruction.

Philosophies of a discipline are ordinarily expected to set forth criteria of belief, and other general characteristics of that discipline, as carried on by ideal, or at least trained, practitioners of it. Instruction in a discipline at any level in school may be viewed as preparation for the appreciation of, and even participation in, that discipline and may be evaluated by its effect in preparing children of a given background for such goals. However, to claim that the general characteristics of a discipline should be found in each or even most instances of school instruction is to make assumptions about the processes of teaching and learning which are not ordinarily implied by philosophical characterizations of ideal or trained disciplinarians. To be sure, there are many useful connections to be made between philosophy and education. For example, Wittenberg [37] presents an interesting and entirely legitimate treatment of the epistemology of mathematics for teachers. Also, some epistemologies, it should be noted, contain explicit statements about psychological processes (see especially Piaget's genetic epistemology, Piaget [25, 26], and Grize [14]).

The present controversy, however, has not made its pedagogical or psychological hypotheses explicit. Yet no dispute has arisen over the explicitly expressed objectives of developing problem-solving ability and preparing for appreciation of and participation in the discipline of mathematics. I am led to the conclusion that the differences between the disputants concern implicit pedagogical hypotheses which are subject to empirical investigation rather than objectives or tastes which might not be so open to investigation. In order to focus more clearly the two sets of hypotheses that I find implicit in the controversy, I shall examine the proposals of the two groups from the point of view of an intermediary position which I propose as an improvement on both positions. Naturally, my proposal is as speculative as theirs, perhaps even more so, since it has not yet been tried *in toto* but only borrows from the experience of each group. However, it will serve to formulate the differences in a manner more amenable to experimental inquiry than has been the case to date.

My proposal makes use of a relationship between the philosophical consideration of scientific knowledge and the educational consideration of the teaching of it which appears entirely justified. While epistemological studies do not ordinarily determine how knowledge should be taught, they provide conceptual frameworks which have some *prima facie* value in formulating educational problems for empirical investigation. Some such relevant framework is especially needed in the absence of a theory of learning adequate to deal with these pedagogical problems.

8. *A new proposal for the teaching of mathematics*

In introducing my proposal, in which I attempt to combine the most plausible hypotheses of the reformers and those of their critics, I wish to recall something well known from the investigations of axiom systems. A completely formulated theory can be regarded as involving three parts which may be roughly characterized as follows:

1. A suitable theory of inference, i.e. a set of rules governing the syntax of the theory proper and the chains of inferences that are acceptable as complete proofs of theorems.
2. The theory proper, i.e. a set of statements including the axioms and other theorems usually associated with the theory, together with such additional axioms as would be required to fill the logical gaps left in conventional proofs.
3. One or more interpretations of the theory proper, such as provided by semantic rules, or rules of correspondence between the theory and a model.

These three parts, or levels, represent a formal way of looking at theories that says nothing about the dynamics of theory development. However, with these distinctions in mind, we may now ask about relations between these three levels having to do with theory development in school mathematics instruction.

First, let us note that any level may be treated informally as well as formally. For example, correspondences between elements of a theory and of a model may be pointed out by ostensive definitions instead of by written rules. At another level, syntactical patterns in proofs may be pointed out by what has come to be known commonly (among some American mathematics teachers, at least) as 'arm waving'. Pointing to the parts of statements which can be replaced by parts of other statements is much easier than constructing elaborate 'tree-diagrams' which show each application of a logical principle in a proof. Of course, as the critics have made quite clear, the theory proper can be informal. However, we may ask whether it is helpful in discovering principles of inference, especially informally, if the statements of a theory proper have been written down in a consistent notation and examined for patterns formed by the resulting inscriptions. Likewise it would be interesting to find out if it would be easier to write down statements in the theory proper from an intuitive acquaintance with a model of the theory if one paid some attention to rules of correspondence between model and theory.

The question just raised concerning the teaching of a theory-of-inference is relatively new in school mathematics, perhaps owing its origin to the suggestions of Hendrix [15], and Beberman and Vaughan [4], but the question about correspondence rules is well known to teachers in connection with the notorious difficulties of 'word problems'. Translating from an applied setting into mathematical language is, for the beginning pupil, the most difficult part of solving problems which are written out in terms of familiar everyday situations.

My proposal hangs on the hypothesis that there exists a kind of pedagogical similarity in the relationships between the two successive pairs of levels. That is, I suppose that an interpretation is related pedagogically to its theory proper much as a theory proper is related to its theory of inference. I hypothesize that in teaching particular propositions of a theory, or of a theory of inference, the primary difficulties of the students and the main task of the teacher lie at the level below. However, in developing a coherent and generalizable theory (theory proper), I hypothesize that the primary difficulties lie in the level above, in the theory of inference. Thus a tendency to omit either the bottom level, as is often done in the textbooks of the reformers, or to omit the top level, as is implied in the critics' recommendations, would, I believe, limit the student's ability to solve new kinds of mathematics problems. In particular, it would limit the student's sources of clues that could be useful in problem solving. I propose therefore that, in school mathematics instruction, all three levels be utilized in order maximally to develop both physical and logical intuitions appropriate to the mathematical theory to be taught and its uses.

Now much of the critics' abhorrence of formal proofs probably stems from their tendency to view the inclusion of logic in the school curriculum either as playing a meaningless symbol-manipulation game or as conducting a very subtle investigation in metamathematics. The difficulty of conceiving of a pedagogically useful and simple procedure involving logical principles probably stems partly from a confusion of the epistemological character of each of the three levels with its pedagogical character. For example, to generalize Piaget's terminology slightly (Inhelder and Piaget [16], pp. 272 ff.) one may refer to operations on a physical model (whether it is actually present in the classroom or not) as operations at the *concrete* level, and one may refer to logical arguments constructed from theoretical statements (in conformity with a theory of inference) as operations at the *formal* level.

However, as Beberman's account of pattern learning in children illustrates (Beberman [3]) there is a great deal of similarity pedagogically between the hooking together of open sentences to form a test pattern for a theorem in algebra and the recognition of patterns in arrays of objects which unquestionably lie at the *concrete* level. The fact that such open sentences may have been given one or more physical or abstract interpretations does not obviate the fact that it is the physical arrangements of symbols on the page or chalkboard that is recognized by the child as conforming or not conforming to a given pattern – the pattern of a logical principle. A test pattern is a chain of patterns which, though concrete in nature, have the possibility of use in an argument at the formal level. Thus, language can be examined as an object in its own right without giving up the ability to go back to that which it talks about. Pedagogically, learning to use language requires mastery of transformations of the concrete forms of language while observing simultaneously the changes in what the language says.

Some aspects of proof *are* related to Piaget's criterion of the formal level, namely that the child use a structured set of all possible combinations of properties in a given situation to solve a problem (Inhelder and Piaget [16], pp. 278 ff.). For example, generalizing the conclusion of a proof for the entire domain of the variables, on the basis of a test pattern, involves the notion of the complete domain of the variables, and possibly uniqueness of the identity element for each operation, in reaching the conclusion that any numeral could replace any variable throughout without changing the pattern and hence without changing the consequence relation between premises and conclusion.

Actually, the proposal to combine all three levels in mathematics instruction in order to enhance comprehension is not really new, except as a general plan to permeate a whole course. Teachers using texts developed by several of the new projects have demonstrated, in isolated instances, that one can develop a bit of mathematics rather smoothly making use of all three levels. In such cases, the students are probably not aware that they are operating on three different levels.

9. *Pedagogical hypotheses worthy of investigation*

The critics have made the assumption that there is no value in hooking together strings of sentences in what logicians would recognize as a proof of a theorem, if the theorem in question is already as obvious to the child as the axioms he uses in constructing the 'proof'. However, as Hendrix [15] has pointed out, there is no reason, *a priori*, for a child to trust deductive procedures, and there may be no other way a child can *learn* to trust deductive procedures except by discovering that they reliably serve to connect what are believed connected on other grounds. That is, ideally, they confirm a syntactical connection between statements which intuitively seem to be related and which are also all known to be true. However, I think the critics are correct in asserting that, if the intuitive basis is missing for statements which are being manipulated according to logical rules, then the learning process becomes a game of dubious educational value. That such a game may nevertheless be interesting to children and adults can be learned by actual experience with one known as WFF'N PROOF, devised by Layman Allen [2].

The controversy thus reduces to the question whether there is any general value in acquiring concrete operational facility with principles of logic, when selected for their applicability in constructing complete proofs, or whether such principles only have a curiosity value in school mathematics because they serve to make clear the difference between rigorous logical reasoning and customary mathematical reasoning which may sometimes be said to be rigorous. It is then a question for educational research, and one of considerable interest, whether carefully designed instruction in writing proofs in beginning algebra contributes to the ability of students to solve algebraic equations because of the logical intuition it develops or whether such algebraic skill is better acquired by instruction designed to develop physical intuition, or whether, as I hypothesize, procedures uniting both approaches will achieve better results than either alone. Similar experimental questions arise connected with the writing of complete proofs of simple theorems in geometry, the intuitive development of geometric theory and the student's ability to prove more complex theorems later on.

One might be tempted to overgeneralize my proposal and suppose that programs of instruction in a mathematical theory which combine the development of physical intuition with the development of logical intuition will provide maximally for pupil learning. It should be pointed out, however, that the three-level analysis is over-simplified with respect to many mathematical theories. Thus, except in quite limited theories, proofs often depend on extra-syntactical principles. They may, for example, depend on restrictions of the domain of the variables or on the inspection of a set-theoretic model in order to observe the effect of a given transformation. Intuitions of quite different kinds may be involved in using different models of a given mathematical theory. (Furthermore, the critics have pointed to another whole aspect of the problem, namely, the conscious use of heuristic rules which may add much more to the effectiveness of instruction.)

Mathematical thought is typically a complex of intuition (of both concrete and abstract forms) and deduction (of both syntactical and semantic forms), and it would be foolish to attempt to prescribe one universal formula for teaching it. Instead, the challenge and delight in teaching mathematics is to be found in the artful design of sequences of tasks which lead the student to even greater power of thought. No restrictions are appropriate except as a matter of permitting the student momentarily to focus on an aspect of the total structure of thought that intrigues him and that will enhance his later opportunities to learn. It is in the focussing of learning on particular aspects of a mathematical theory that the reformers and the critics make different contributions to mathematical pedagogy. But focus can shift about, so the two sets of recommendations should not be regarded as mutually exclusive.

At this point it is helpful to note the advice of Professor Dienes [10] that mastery of any mathematical concept is acquired more surely if the student has the opportunity to work with multiple embodiments of that structure – especially if they are of rather diverse kinds. This notion is beginning to come into the reform projects. For example, Professor Rosenbloom is writing a series of elementary school texts which present mathematical ideas in the contexts of a variety of science experiments. Davis [8] has experimented with presenting a given mathematical idea in a variety of ways, including both geometric diagrams and imaginary social situations. Educators are also beginning to take note of such work as that of Inhelder and Piaget [16] who have pointed out numerous embodiments of certain logical operations. While it is entirely possible and even pedagogically feasible to employ multiple embodiments of the particular group of logical principles which are most used in mathematical proofs, these would certainly tend to be written arguments of one kind or another. Such embodiments lack the ease of manipulability and the responsive or confirming character possessed by semi-concrete devices such as Dienes employs.

It was the search for some manipulative device that would behave in accord with the logical principles of proof that led me to undertake the use of a computer to check steps in proofs as they are being constructed on line by a student. As I later discovered, Suppes has also developed a system permitting construction and checking of truth tables and proofs of logical theorems (such as those found in Suppes and Hill, *First Course in Mathematical Logic* [34]). At Illinois, Gelder, Golden, Smith, and I are now generalizing a computer program we had designed earlier for the computer-based teaching system that goes by the acronym of PLATO (Bitzer and Easley [7]). The generalized program will handle proofs in logic and elementary goemetry as well as in algebra (Easley *et al.* [11]).

What is of interest here is that our program is designed to permit the student to invent his own version of principles of logic in order to account for the machine's behavior in a way that is plausible to him. Because the student must specify fully the basis for each step, i.e. the substitution he wishes to make in the axioms of his own choice and where in a previous line of proof the resulting instance of his axiom is to be applied, mastery of the principles of substitution and replacement is assured by successful manipulation of the device. Replacement by the right member of the axiom instance is carried out for the student only if the left member matches the string of symbols beginning where the student indicated, otherwise not. Instances of *modus ponens* are generated in a similar way. Experimentation with logical organization of a theory is made possible because the student can insert his own axioms and lemmas and make use of any theorems already proved in composing new proofs. It is even possible in principle to require less specificity from the student. For example, after he has mastered substitution, replacement, and modus ponens, these operations might be made automatic in most instances.

Computer programers and engineering designers report that they learn faster by having a computer check the program they have written for consistency or check their design for performance, than was possible by lively and critical discourse alone. Similarly, learning in the classroom may be enhanced by the installation of computer consoles which permit teachers to program a computer to respond in accord with a model of their choice, and at which children can get their homework checked, compose proofs, or check hypotheses about a model of current interest – all as individualized supplements to classroom discourse. Also, in case a real computer isn't available, the program *itself* can often be simulated by teachers or fellow students. Children who enjoy complete proofs in algebra are very likely to enjoy composing and checking computer programs by hand, and we should also find out whether this activity is an aid to developing logical intuition.

There may be a tendency to regard heuristic procedures as

necessarily associated with the use of physical models and informal proofs. Pólya's impressive work on heuristic procedures in problem solving and his application of them to informal mathematics teaching has, for the most part, been limited to geometric and arithmetic systems. This tendency may account in part for the general failure of the reform projects to recognize the general applicability of heuristic rules. There is, however, good reason to believe that heuristic procedures apply equally well to the manipulation of sentences in formal proofs. For example, Newell *et al.* [23] have demonstrated several heuristic rules that apply to the generation of complete proofs. Their ends-means analysis and other heuristic rules, when programed on a computer, are reported to have successfully proved 70 % of the theorems in Chapter II of *Principia Mathematica*, and simulated human performance in the process (see Newell and Simon [24]). Such heuristic rules should be applicable in classes where students are proving theorems in the theory proper as they investigate logic informally.

In a very different vein, Lakatos has extended Pólya's heuristic procedures to a theory of proof analysis in an essay intended to show that 'informal, quasi-empirical, mathematics does not grow through a monotonous increase of the number of indubitably established theorems' but by 'the logic of proofs and refutations'. His demonstration of how proofs and counterexamples may serve jointly to improve conjectures and improve upon initial, informal proofs – thus moving in the direction of a useful axiomatization by exploration of relationships between conjectures – suggests a powerful pedagogical technique which clearly has value in teaching topics from mathematics proper. But as Lakatos observes, the same patterns of growth are to be found in the recent history of metamathematics ([20], p. 6). This observation suggests that the 'logic of proofs and refutations' could also be useful in the informal, classroom development of a theory of inference. I expect that a skilful teacher of algebra or geometry could certainly make use of this notion in developing formal criteria of proof, using proofs accepted on intuitive grounds as raw material for studying the consequence relation.

It is apparent from the literature about and the textbooks of the new mathematics programs that both reformers and critics are primarily committed to the development of problem-solving ability. Yet reflection on the critics' objections to formal proof suggests a still more basic concern in mathematics education. It appears that what the critics object to most in the new programs is the tendency they have to rely on logical principles in deciding what are and what are not consequences of given axioms. In this the critics may be making a complaint about the authoritarian position often taken by the new textbooks especially on principles of logic, but also in specifying what the axioms of a theory are to be. This indeed is a legitimate complaint philosophically, and a point which deserves serious consideration. Since both the critics and the reformers have emphatically subscribed to the objective of making the student a skilful mathematical inquirer, this would seem to imply that both would favor the teaching of principles of rational inquiry, at least implicitly.

In teaching for reasoned evaluation of knowledge, as Scheffler [33] has pointed out, practice in the discipline of reasoning must be provided the student. Scheffler observes that informing the student of certain facts does not thereby guarantee that he knows these facts – knowledge requires a reasoned assurance of truth which mere information does not. The autonomy of the knower must be evidenced, Scheffler argues, 'in the ability to construct and evaluate fresh and alternative arguments, the power to innovate, rather than just the capacity to reproduce stale arguments earlier stored' ([33], p. 140). This requires mastery of principles of reasoning including 'rules of inference and procedure in the special sciences', ([33], p. 141). In mathematics, these principles surely include both the logic of proof and heuristic procedures demonstrably relevant to problems the student is expected to solve.

Principles of inquiry treated in a mathematics class ought to be directly appropriate to the inquiry at hand and ought not to set off an infinite regress of justification upon justification. We find a guide to limiting authoritarianism in Pólya's writing, when he argues that each step in a lesson should be of such a character that the student might have thought of it himself. What Pólya apparently fails to recognize is that the student might very well think of (i.e. notice) the relationships of patterns in written statements and ask if such relationships are trustworthy guides to relationships in the model to which the statements refer.

It is my contention that authoritarianism is far more easily remedied by the procedure I have suggested than by the procedures which the critics have suggested. The reformers have produced a detailed logical analysis, in which the principal student questions that have arisen during years of classroom tryout are answered, and this should not be discarded as a framework without clear evidence that it can be superseded by some other. Furthermore, a well-organized theory of inference is available as support for the teacher who encourages the search for logical principles by his pupils. With this material in hand, it is also possible to train teachers for the task of leading students in a non-authoritarian manner through such a development and to provide them with adequate practice in the needed elements of skill.

10. *Conclusions*

We can now see the educational mistake in either defending or criticizing instructional programs in terms of their conformity or non-conformity with given epistemological views of the nature of mathematics. From the point of view of teaching, what appears to be a formal treatment of elementary algebra may, for the student, be an informal treatment of principles of logic. What appears to be a formal treatment of relations and functions may, for the student, be an informal treatment of the correspondence between algebraic statements and a set-theoretic model. The pedagogical hypothesis that emerges is that informality is required on the level at which relationships are unfamiliar to the student, in order to avoid either losing him on the consequence relation, a practice for which mathematicians are notorious, or losing him on the correspondence between physical model and theory, which is a problem well known to teachers. (That the latter is also a problem of growing concern to scientists and engineers with respect to the curricula of colleges and universities is evidenced, for example, by Weinberg's recent criticism, [35]). The new programs attack the former problem, but in a more authoritarian way than necessary. The critics' recommendation that instruction always begin with a familiar model and employ heuristic rules attacks the latter problem, but in a more limited way than necessary.

In our proposed intermediate procedure, which utilized the conceptual framework of formalist investigations but not formalist epistemology, a somewhat generalized version of the critics' pedagogical principle is preserved – namely, there should always be an aspect of instruction which is open, which is treated informally, in which principles are guessed at from a model familiar to the student, and in which heuristic rules can be introduced for problem solving. At the same time, by extending informality to a theory of inference, and, by extending the use of models to include chains of sentences and structured sets, the contributions of the reformers to clarifying the consequence relation and to the development of logical intuition can be retained.

Since some of the reformers have, on occasion, successfully employed this intermediate procedure (except for its heuristic aspect), it seems plausible that, fully applied, it may achieve the objectives of the reformers and critics, but without following the latter's advice in completely removing formal treatment of mathematical topics. The plausibility of developing logical principles pertinent to the theory at hand stems from the notion that, if the coherence of the various propositions of a theory is to be grasped, one must also learn the principles of inference that permit linking them. The fact that both correspondence rules and principles of inference are frequently omitted in traditional mathematics instruction is plausibly a principal reason why many students have had difficulty in making sense of school mathematics, even with the best of teachers.

The reformers, despite their excellent work in organizing mathematics for teaching, have tended to present mathematics in an authoritarian manner. They have too often introduced axioms and logical rules without developing them as plausible representations of known facts about a model and have generally over-looked the possible pedagogical aid that heuristic rules might add. However, they *have* developed pedagogical techniques for helping pupils discover patterns, and, if these techniques were used more extensively, both with models and with logical principles, they would remove the grounds for the imputation of authoritarianism, a philosophically defensible criticism I find implicit in much of what the critics have said.

From the critics we may also learn that the really new venture for mathematics curriculum reform lies in the area of heuristic procedures. If mathematics educators learn to apply the insights into mathematical inquiry which Polya and Lakatos have set forth, at the level in the teaching of mathematics on which the growing edge of the student's understanding happens to lie, the interest and achievement of students may be expected to increase markedly. And, with such resources being applied to curriculum design, we may also hope to enhance the intellectual appeal of the educational enterprise. We must continue to attract to the field of education more and more persons talented in the sciences, mathematics, logic, philosophy, and pedagogy. For to such talent we must look for an improvement in the quality of education sufficient for an age in which hope hangs heavy on the rapid spread of intelligence and intellectual skills.

References

[1] Lars V. Ahlfors and 64 other mathematicians, On the mathematics curriculum of the high school, *American Mathematical Monthly* **69**, (1962), pp. 189–193, also published, in the same year, in *The Mathematics Teacher* **55**, pp. 191–195.

[2] Layman E. Allen, *Wff'n proof, The Game of Modern Logic*, Wff'n proof, New Haven, Conn. (1962).

[3] Max Beberman, Searching for patterns, in *Mathematics Today; A Guide for Teachers*, Howard F. Fehr, ed., Organization for Economic Co-operation and Development, Paris. (1963).

[4] Max Beberman and Herbert E. Vaughan, *High School Mathematics, Unit* 6, *Teacher's Edition*, Urbana: University of Illinois Press. (1960).

[5] Max Beberman and Herbert E. Vaughan, *High School Mathematics, Course* 1, *Teacher's Edition*, Boston: D. C. Heath and Company. (1964).

[6] E. G. Begle, Some remarks on 'On the mathematics curriculum of the high school", *The Mathematics Teacher* **55** (1962), pp. 195–196.

[7] Donald L. Bitzer and J. A. Easley Jr., plato: a computer-controlled teaching system, in Sass, Margo A. and Wilkinson, William D. (eds.) *Computer Augmentation of Human Reasoning*, Washington: Spartan Books, Inc., (1965), pp. 89–103.

[8] Robert B. Davis, Goals for school mathematics: The Madison Project view, *Journal for Research in Science Teaching* 2 (1964), pp. 309–315.

[9] Benjamin De Mott, The math wars, in *Hells and Benefits*, by the same author, Basic Books; originally published in *The American Scholar*, also in Robert W. Heath, ed., *New Curricula*, Harper and Row, 1964. (1962).

[10] Z. P. Dienes, *An Experimental Study of Mathematics Learning*, London: Hutchinson. (1963).

[11] J. A. Easley, Jr., H. M. Gelder and W. M. Golden, A plato program for instruction and data collection in mathematical problem solving, Coordinated Sciences Laboratory Report R–185, University of Illinois, Urbana. (1964).

[12] Dorothy Friedman, editor, *The Role of Applications in a Secondary School Mathematics Curriculum*, Proceedings of a UICSM conference held at Allerton House, Monticello, Illinois, February 14–19, 1963; UICSM, 1210 W. Springfield, Urbana. (1964).

[13] John I. Goodlad, *School Curriculum Reform*, New York: The Fund for the Advancement of Education. (1964).

[14] Jean-Blaise Grize, Genetic epistemology and psychology, in Wolman, B. B. and Nagel, E. (eds.) *Scientific Psychology*. New York: Basic Books, (1965), pp. 460–473.

[15] Gertrude Hendrix, The psychological appeal of deductive proof, *The Mathematics Teacher* 54 (1961), pp. 515–520.

[16] Bärbel Inhelder and Jean Piaget, *The Growth of Logical Thinking from Childhood to Adolescence*, New York: Basic Books. (1958).

[17] Morris Kline, *Mathematics in Western Culture*, New York: Oxford University Press. (1953).

[18] Morris Kline, *Mathematics and the Physical World*, New York: Thomas Y. Crowell Company. (1959).

[19] Morris Kline, *Mathematics, A Cultural Approach*, Reading, Mass: Addison–Wesley Press. (1962).

[20] Imre Lakatos, Infinite regress and the foundations of mathematics, *The Aristotelian Society Supplementary Volume* 36 (1962), pp. 155–184.

[21] Imre Lakatos, Proofs and refutations, *The British Journal for the Philosophy of Science*, 14 (1963), pp. 1–25, 120–139, 221–245, 296–342.

[22] Edwin E. Moise, Alexander Calandra, Robert B. Davis, Morris Kline and Harold M. Bacon, Five views of the 'new math', *Council for Basic Education Occasional Papers, No.* 8, Washington, D.C. (1965).

[23] A. Newell, J. C. Shaw and H. A. Simon, Empirical explorations of the Logic Theory Machine, *Proceedings of the* 1957 *Western Joint Computer Conference*, (1957), pp. 218–230.

[24] A. Newell and H. A. Simon, Computer simulation of human thinking, *Science* **134** (1961), pp. 2011–2017.

[25] Jean Piaget, Epistémologie Mathématique et Psycologie, deuxiéme partie, *Etudes d'Epistémologie Génétique*, xiv, Paris: Presses Universitaires de France. (1961).

[26] Jean Piaget, Psychology and philosophy, in Wolman, B. B. and Nagel, E. (eds.) *Scientific Psychology*, New York: Basic Books, (1965), pp. 28–43.

[27] George Pólya, *How to Solve It*, Princeton: Princeton University Press. (1945).

[28] George Pólya, *Mathematics and Plausible Reasoning* (2 vols.), Princeton: Princeton University Press. (1954).

[29] George Pólya, The teaching of mathematics and the Biogenetic Law, in I. J. Good, ed., *The Scientist Speculates*, London: Heinemann. (1962a).

[30] George Pólya, *Mathematical Discovery*, Vol. 1, New York: John Wiley & Sons, Inc. (1962b).

[31] George Pólya, *Mathematical Methods in Science* (Leon Bowden, Ed.), Volume xi of *Studies in Mathematics*. Stanford, California: School Mathematics Study Group. (1963).

[32] George Pólya, *Mathematical Discovery, On Understanding, Learning and Teaching Problem Solving*, Vol. ii, New York: John Wiley & Sons, Inc. (1965).

[33] Israel Scheffler, Philosophical models of teaching, *Harvard Educational Review* **35** (1965), pp. 131–143.

[34] Patrick Suppes and Shirley Hill, *First Course in Mathematical Logic*, New York: Blaisdell Publishing Company. (1964).

[35] Alvin M. Weinberg, But is the teacher also a citizen? *Science* **149** (1965), pp. 601–606.

[36] Alexander I. Wittenberg, *Vom Denken in Begriffen*, Basel and Stuttgart: Birkhauser. (1957).

[37] Alexander Wittenberg, An unusual course for future teachers of mathematics. *American Mathematical Monthly*. **70** (1963a), pp. 1091–1097.

[38] Alexander I. Wittenberg, *Bildung und Mathematik – Mathematik als exemplarisches Gymnasialfach*, Klett, 1963. Stuttgart: Klett. (1963b).

[39] Alexander I. Wittenberg and Sœur Sainte Jeanne-de-France, *Redécouvir les Mathematiques*, Neuchâtel. (1963).

[40] William Wooten, *SMSG, The Making of a Curriculum*, New Haven: Yale University Press. (1965).

DISCUSSION

P. Suppes: *The central role of empirical knowledge in curriculum reform.*

The point of my comments is not to disagree with Easley but to amplify and perhaps give stronger emphasis to some of the points central to his paper. The general aspect of curriculum reform that would seem to have been too little considered by reformers and critics alike is the acquisition of empirical knowledge as a basis for change. At times it would seem that the definiteness and fervour with which views are expressed are inversely related to our state of knowledge of the subject. What is perhaps most surprising is that mathematicians and other scientists, who are volubly critical of the shallowness of traditional educational research, have begun the development of a body of educational literature that is even shallower and, in many cases, fundamentally anti-scientific and anti-intellectual in spirit and substance.

What I want to urge is the central relevance of empirical knowledge of several kinds for changing or criticizing changes in curriculum. There are at least four sorts of empirical investigations that are essential. One sort is relatively straightforward data about the educational background of teachers and their actual knowledge of subject matter, whether traditional or modern. The results of tests given over several decades on the traditional mathematical knowledge of elementary-school teachers in the United States, for instance, have indicated how incomplete is the mastery by many teachers of even the concepts of ordinary arithmetic.

A second sort of investigation is concerned with the collection and analysis of standardized test data to evaluate and compare the performance of students learning a subject in a traditional or modern curriculum guise. There are inherent methodological difficulties in comparative evaluative investigations of this sort, and yet they are a necessary part of curriculum reform. In the case of mathematics, a fairly considerable body of test results have now been published in the United States, but it is characteristic of the intellectual tenor of the critics whom Easley discusses that they seldom if at all enter into a detailed analysis of the test results that have been published.

A third sort of empirical investigation attempts to move to a deeper scientific level by analysing in detail the learning, and particularly the learning difficulties, of students as they work their way through a given curriculum. Referring again to the case of mathematics education, it is disappointing that the good scientific beginning in these matters which was due to the pioneering efforts of such investigators as Thorndike and Brownell in the twenties and thirties has not deepened and become a genuine body of science. Practically any sophisticated question we can ask about the learning of elementary arithmetic has not yet received a fully satisfactory scientific answer. To mention one example, the number of systematic studies of response latencies in the training or maintenance of arithmetic skills can almost be counted on the fingers of one hand, and yet there is good reason to think that latency data provide the most sensitive practical measure of item difficulty, and one of the few observable indices of the 'mental' steps involved in problem solving.

A fourth sort of empirical investigation is concerned with scientific problems that transcend curriculum issues and that are central to the development of an empirically adequate theory of complex learning and concept formation. It is probably unfortunately true that some of the deeper issues mentioned by Easley as dividing reformers and critics will not be satisfactorily settled outside the framework of a sophisticated systematic theory of learning which is empirically correct to a first approximation. There is some hope, I think, that current research in mathematical psychology and in computer science – particularly the theory of artificial intelligence – is on the verge of discovering learning theories that will move us a definite step closer to constructing this framework.

In conclusion let me say that I certainly recognize the necessity of moving ahead with curriculum reform and with continuing the dialogue between reformers and critics, without waiting for an adequate body of empirical knowledge to be accumulated. My point in these comments has been rather to emphasize the continuing importance of accumulating knowledge as the process of change goes on, and to urge reformers and critics alike to use to the maximum what knowledge is available.

L. KALMÁR: *Mathematics teaching experiments in Hungary* [1].

You may be interested, in particular from the points of view raised by Professor Easley, in how mathematics is being taught in Hungary – a country which has given the world far more outstanding mathematicians, compared with the number of its inhabitants, than any other country.

In Hungary, the most current secondary curriculum (broadly, from 14 to 18) includes elementary algebraic identities, linear and quadratic equations, systems of linear equations with two (exceptionally with three) unknowns, simple functions of one variable, powers, roots and logarithms, arithmetical and geometrical progressions, elementary Euclidean geometry (plane and solid), trigonometry and elementary analytical geometry (of the plane); and it will soon contain (mainly at the initiative of physicists) elementary calculus and probability. As in any other country, an overwhelming majority of schoolboys (and schoolgirls) find mathematics, taught by traditional methods on the basis of this traditional curriculum, awfully boring and extremely difficult to learn. And even when it is learnt, by extremely strenuous work and exhaustive use of one's diligence, it is found to be useless in problem solving, which is, even in the case of standard school mathematical problems, quite another art, and one almost impossible to master. Nevertheless, now and again we find in many schools pupils who get the textbook in their hands, read it, find it very interesting, and acquire its content, including its application to problem solving, within a few days, and are full of keen desire to learn more. They do not understand why and how their fellow-pupils find difficult what they read with interest and found obvious. They also are awfully bored by the teaching in the school, the teaspoon doses

[1] I am indebted to Mr. T. Varga for valuable information which I used in the final formulation of this contribution.

for their fellow-pupils of what they know already. The discrepancy between the 'average pupils' and those who are generally called 'talented in mathematics' is so sharp (though there are a few transitional cases) that one is inclined to regard the latter as mutants of normal *homo sapiens.*

Up to 1962, the only possible, though very effective, method of turning pupils talented in mathematics into working mathematicians was to treat them individually, to satisfy their curiosity by describing to them branches of mathematics they did not yet know, and giving them problems to test their knowledge and problem-solving ability. Individual treatment was supplemented by different degrees of mathematical problem solving competitions, one of which, organized each year for students finishing their schools in that year, is now called the Kürschak competition. This competition, which has the oldest tradition of all, is well known throughout the mathematical world from the Hungarian Problem Book, which contains the problems proposed for these competitions, with their solutions.

Though professional mathematicians often proposed to extend and improve the curriculum by including interesting modern subjects, such as graph theory, mathematical logic, set theory, and the like, eventually the reform of the curriculum became restricted to such questions as the omission or admission of calculus. And in spite of such efforts to reshape teaching methods as the publication of an excellent series of textbooks by T. Gallai, R. Péter (*et al.*) the majority of teachers stuck to the old 'well-tested' methods. Again, in spite of repeated suggestions by the Bolyai János Mathematical Society that special experimental classes should be formed where pupils gifted in mathematics could learn at their own pace, hence much more effectively than average pupils do, the dogmatic, pedagogical views of those who feared that talented pupils would be spoiled if they became conscious of their talents prevented the realization of such plans before autumn 1962.

The International Symposium on School Mathematics Teaching, organized in 1962 by the Hungarian National Committee for UNESCO and the Ministry of Culture of Hungary with the collaboration of the Bolyai János Mathematical Society and with assistance from UNESCO, was decisive for mathematics teaching reform in Hungary. The Conclusions and Recommendations of the Symposium encouraged those who sympathized with endeavours to reform mathematics teaching, and the decision was: let us try! Soon an experimental 'mathematical class' was organized in the Fazekas Secondary School in Budapest and later in three other schools (two in Budapest and one in Debrecen). These schools select their pupils for the mathematical classes from an increasing number of applicants.

The experimental curriculum includes, besides the usual subjects (see above), calculus, probability and statistics, numerical methods in equations, numerical integration, elements of nomography, graph theory, sets and logic, and computer programming; and it gives the teacher a rather free hand in the details of his work. Hence it is possible for the pupils themselves to modify the subject-matter by their questions.

As to teaching methods, first I must explain that in our country, as opposed to the United States, the 'reformers' and the 'critics' in the sense of Mr. Easley's paper, are, curiously enough, the same. Hungarian mathematicians who are enthusiastic about mathematics curriculum reform are at the same time convinced that trying to teach arbitrary rules, or abstract concepts without sufficient preparation in more concrete particular cases, or even without showing how general concepts are formed by abstraction, does not serve any rational purpose. They are convinced that mathematical rigour can be acquired not by imitating a teacher who simply adopts and applies the present level of rigour without explaining how the science reached it, but only by developing the pupil's taste for rigour by starting with the intuitive point of view and showing repeatedly why some degree of rigour becomes necessary for certain problems. So we, enthusiasts of reform in mathematics teaching, instead of fighting 'Math Wars' against each other as they do in the United States, can unite our forces against formalism in teaching, represented by many conservative teachers.

Hence, in mathematical classes, much stress is laid on heuristics. The teacher starts, in general, with a problem which seems interesting for the pupils and develops the necessary concepts and methods with the motivation that they are useful tools for solving the problem in question (and several other problems, some of a quite different character).

What are the results?

First, pupils of mathematical classes learn better and feel much happier than pupils, talented in mathematics, attending ordinary classes. They never lose their interest. To tell the truth, many of them are more interested in theoretical questions of mathematics than in its applications; e.g. about half the pupils of the first mathematical class want to study mathematics after leaving school, in order to become research mathematicians, the rest wish to become teachers, engineers, economists and physicians. Though much more stress is laid in the experimental classes on solving problems than on a formal presentation of mathematical theories, some of the pupils can present their thoughts, even in written form, quite clearly, while others have, of course, the usual difficulties of the beginner in this respect.

Such are our results in teaching pupils talented in mathematics. And what about the rest? Do we accept as an unchangeable fact that they are lost for ever to the beauties (hence also to the useful applications) of mathematics, that they can learn only dull rules for solving simple computational problems of standard types, without being able to understand why (and when) these rules work?

We must admit that our efforts to raise the average level in mathematics of average pupils in secondary schools has failed so far, or at least we do not have considerable success in this field. The main reason may be that pupils, having been spoiled by dull elementary teaching, lose their ability to understand (and be interested in) mathematics.

Such considerations led an enthusiastic mathematician, Mr. T. Varga, to initiate in autumn 1963 – after a series of preliminary experiments, begun in 1957 – a mathematics teaching experiment in two parallel classes of an elementary school in Budapest, starting with children at the age of six. The pupils of one of the classes had not been selected at all; the other was a 'musical class', whose pupils had been selected for their musical talent. The curriculum was prepared by Mr. Varga on the basis of his former experiences as well as on the results and ideas of Z. P. Dienes, R. B. Davis and others. Two excellent teachers, Mrs. L. Barabás and Mrs. G. Tiszai, taught the pupils in all subjects (except singing in the musical class).

They taught computation with bases different from 10 (using Dienes' Multibase Arithmetic Blocks, from the first class on), functions (defined by tables in the first class, by graphs in the Cartesian coordinate system from the second class on, and by formulae from the third class on), elements of set algebra (by Venn diagrams, from the second class on), binary relations (by arrow diagrams, from the third class on), some graph theoretical concepts, and the like. This they did, of course, not by the deductive method, but starting with games and gathering experiences from the physical world. They taught, further, linear equations and inequalities, systems of linear equations, also with negative solutions (in the third class), as well as quadratic equations (first by trial and error, then on the basis of the relations between roots and coefficients).

The accomplishment of the pupils has been several times tested, both in the traditional (routine arithmetic) and in the additional material. As regards to the first, it has been found that pupils of the experimental classes did considerably better though their time allotted to the traditional material was not more than ono third of the usual. As to the additional material, no comparison was possible with parallel classes which have been restricted to traditional topics. Comparisons with higher grades are contemplated but not yet carried out.

These experiences show that, by appropriate teaching methods, the effectiveness of teaching mathematics can be considerably raised. Perhaps the pupils talented in mathematics are not mutants of *homo sapiens* at all, and to be able to understand, to apply, and to be interested in mathematics, when it is well taught, is a general property of *homo sapiens*.

J. A. Easley, Jr.: *Reply.*

Professor Kalmár's description of mathematics curriculum development in Hungary suggests to me the influence of the same philosophical views concerning the nature of mathematics that have shaped mathematics curriculum recommendations in the United States. Conflict is avoided, apparently, because one viewpoint (the applications-oriented, heuristic approach) is directed to the education of talented secondary school students, and the other (the mathematical systems, logic-oriented approach) is directed at general classes of primary grade pupils. Although empirical evidence, not philosophical arguments about the nature of mathematics, is required to resolve questions concerning the best curriculum or the best instructional practices, it is just possible that the Hungarian mathematics curriculum reformers have intuitively placed these emphases in much more appropriate positions in the curriculum than some of their American counterparts. When children trained in Varga's program meet the new secondary school program, a smooth and non-controversial connection between logic and rigorous mathematical argument should be possible. Kalmár's report is in keeping with my conviction that, at the present stage of our knowledge of complex learning, epistemological views are more appropriately used as sources of ideas for empirical curriculum research than as competing philosophies of mathematics education [1]. We are all indebted to Kalmár for giving us such an interesting account of the Hungarian work.

I am also very much obliged to Professor Suppes for providing an emphasis on the practical problems of empirical research in resolving curriculum issues – a matter I completely neglected in my paper. His observation that the contributions of scientists to curriculum work often give the impression of an anti-intellectual

[1] The reader's attention is called to a collection of papers on the nature of mathematics and mathematics instruction, published too late to receive attention herein. It is entitled, *The Role of Axiomatics and Problem Solving in Mathematics*, by the U. S. Commission on Mathematics Instruction, Boston: Ginn and Company, 1966. Contributors include George Polya, Morris Kline, Patrick Suppes, Herbert E. Vaughan, and Paul C. Rosenbloom.

attitude (because of their neglect of empirical studies) is an important one, and I fear that the impression may have a measure of truth in it. However, I should like to take advantage of this opportunity to suggest some qualifications to his remarks and to recommend still another kind of empirical curriculum research.

Standardized test data concerning mathematical knowledge of students and teachers are quite useful for purposes of detecting weaknesses in course design and in teacher training; however, in my opinion, collections of such data have, so far, not provided an adequate basis for choosing one approach to mathematics instruction over another. One possible exception is found in the data which Suppes and his colleagues have collected concerning pupil achievement with their primary-school arithmetic materials. It may be true, as Suppes remarks, that the critics of 'new math' programs are not willing to examine this sort of data, but I would be hard pressed to locate any test data at the secondary school level that I could recommend as answering significant questions. I would be equally dubious about the research value of problem-solving competitions of the sort Kalmár describes, although these kinds of problems are often recommended for purposes of curriculum evaluation.

Like Suppes, I have been impressed by the potential contribution that could be made through studies of learning and learning difficulties and by the possible foreshadowings of a theory of complex learning. I join with him in urging further work along these lines, but I believe that a more immediate approach to the complex processes of mathematics instruction is both possible and urgently needed. Philosophical analysis has often been related in Socratic fashion to the role of a teacher interacting with his pupil inquirers. A teacher may also interact with his students as a fellow inquirer. The effects of such interactions on interest and learning will be contingent upon many factors, but they can be recorded by well-trained observers and analyzed systematically from some conceptual framework such as, for example, the one I have outlined in my paper.

A teacher and his pupils constitute a system which can hardly be expected to function with maximum effectiveness if the major sources of external influence (textbooks and teacher training programs) only reflect their individual learning characteristics and prior knowledge. Surely such influences need to be designed, above all, in terms of the kinds of interactions that pupils and teachers are able to have with each other and with the subject matter. This requires a sharp departure from traditional research approaches as well as talented teachers who are given a free hand so that pupils have an opportunity to contribute to the shaping of the curriculum – much as Kalmár has described in the Fazekas Secondary School. This procedure may be regarded as analogous to Piaget's studies of the development of children's thought, which turn epistemological problems into psychological ones.

When curriculum reformers and critics have turned deaf ears to the suggestion that empirical research is needed, it is sometimes due to a misinterpretation of the term in the narrow sense of formal experiments and measurements and not because they are anti-empiricists in general. The training of research workers, the funding of research, and the editorial policies of research journals in education in the United States have been so strongly influenced by the models of psychological measurement and inferential statistics that no other interpretation of 'educational research' is readily put into practice. However, in the exploration of pupil-teacher interactions in mathematics instruction, the variables that are easily recognized are too numerous and the hypotheses that suggest themselves to the careful observer shift too quickly for these models to be useful.

Some kind of scientific discipline is required if classroom explorations are to make a cumulative contribution to a rational understanding of mathematics teaching. Mere rejection of the presently popular notion of educational research is indeed anti-intellectual. Perhaps an approach somewhat like that employed in cultural anthropology would be more useful. If observations were collected by a philosophically trained observer who was guided by some general notions of the characteristics of intellectual functioning, there would be a greater opportunity to understand the system as a whole than there is through formal research designs. Such an understanding, aside from its intellectual value, is very badly needed for the improvement of teacher education programs.

A little work of this sort has been started. Hendrix has collected detailed observations in Beberman's classroom, using movies, still photographs, and dialogue transcription. Davis and Page have also employed films in this type of work, but the analysis of such data has not been reported, although it presumably has had an influence on the materials they have produced for teachers and pupils. It is clear that the pressures for production of curriculum materials have been much stronger than scholarly interest in behavioral research. As a consequence, the literature of the reform movement does not permit one to distinguish between those reformers who have been influenced by careful classroom exploration and those whose recommendations lack this kind of empirical control. As a whole, it presents an undisciplined picture (and sometimes even a warlike picture). This situation should be remedied by the joining of serious reformers, critics, mathematics educationists, and philosophers in vigorous research activity and publication along lines they find most appropriate to their concerns – one of the more formal types of research that Suppes recommends or the more exploratory approach I have suggested.